网络专业校企合作开发项目式教学系列教材

企业级网络构建技术实训教程

主　编　邵长文
副主编　李振松　史建政
参　编　白树成　刘学普

电子工业出版社
Publishing House of Electronics Industry
北京·BEIJING

内 容 简 介

本书是网络专业校企合作开发项目式教学系列教材。本书基于"项目导向、任务驱动"的项目化教学方式编写而成，体现"基于工作过程"，"教、学、做"一体化的教学理念。

本书从实际应用出发，为培养学生的实际动手能力，分为交换部分和路由部分，共计 26 个实训项目。从交换机端口操作到 VLAN 间路由，以至生成树原理分析；从 OSPF 路由技术到 BGP 路由技术，以至较高级策略路由和路由策略技术；旨在通过实训项目的安排，使学生能够将课堂中学习到的知识技术在实训中得到训证、加深对相关知识点的理解和技术掌握。本书结合教材中的与实训相关章节内容的学习，提前做好实训预习，做到实训前明确实训目标、掌握实训的基本内容及操作方法；在实训中正确使用实训环境，认真观察实训结果；实训后针对实训目标，认真思考总结，梳理成功与不足，写出实训报告，将知识、技术和能力融会贯通，从而做到学以致用。

本书既可以作为高职院校计算机应用专业和网络技术专业理论与实践一体化教材使用，也可供相关领域的工程技术人员学习、参考。

未经许可，不得以任何方式复制或抄袭本书之部分或全部内容。
版权所有，侵权必究。

图书在版编目（CIP）数据

企业级网络构建技术实训教程 / 邵长文主编．—北京：电子工业出版社，2014.7
网络专业校企合作开发项目式教学系列教材
ISBN 978-7-121-23029-5

Ⅰ．①企⋯ Ⅱ．①邵⋯ Ⅲ．①企业－计算机网络－高等学校－教材 Ⅳ．①TP393.18

中国版本图书馆 CIP 数据核字（2014）第 080414 号

策划编辑：王羽佳
责任编辑：郝黎明
印　　刷：北京七彩京通数码快印有限公司
装　　订：北京七彩京通数码快印有限公司
出版发行：电子工业出版社
　　　　　北京市海淀区万寿路 173 信箱　邮编：100036
开　　本：787×1 092　1/16　印张：11.25　字数：288 千字
版　　次：2014 年 7 月第 1 版
印　　次：2017 年 1 月第 2 次印刷
定　　价：35.00 元

凡所购买电子工业出版社图书有缺损问题，请向购买书店调换。若书店售缺，请与本社发行部联系，联系及邮购电话：(010)88254888。
质量投诉请发邮件至 zlts@phei.com.cn，盗版侵权举报请发邮件至 dbqq@phei.com.cn。
服务热线：(010)88258888。

前　言

　　企业级网络构建技术实训教程为适应计算机网络技术专业技能培养的要求，根据实际工作过程所需的知识和技能抽象出若干个实训教学项目，形成了高职计算机网络技术专业学生量身定做的网络工程实训教材。本书从职业的岗位分析入手开展教学内容，强化学生技能训练，在训练过程中巩固所学的知识。

　　该教材有如下特色：

　　体现"项目导向、任务驱动"的教学特点。

　　从实际应用出发，从工作过程出发，从项目出发，采用"项目导向、任务驱动"的方式，通过"项目提出"、"项目分析"、"项目实施"三部曲展开教学。在教学设计上，以工作过程为参照系来组织和讲解知识，培养学生的职业技能和职业素养。

　　体现"教、学、做"一体化的教学理念。

　　以学到实际技能、提高职业能力为出发点，以"做"为中心，教和学都围绕着做，在学中做，在做中学，从而完成知识学习，技能训练和提高职业素养的目标。

　　本书体例采用项目案例形式。

　　全书设有 26 个项目案例，教学内容安排由易到难、由简单到复杂，循序渐进。学生能够通过项目学习，完成相关知识的学习和技能的训练。

　　项目案例的内容体现典型性、实用性、趣味性和可操作性。

　　本书力求体现教材的典型性、实用性、趣味性和可操作性。根据职业教育的特点，针对中小型网络实际应用，编写 Linux 网络操作系统课程的实用型教材。减少枯燥难懂的理论，重点对网络服务的搭建、配置与管理进行全面细致的讲解，理论联系实际多一些，突出工程实践案例的实训。

　　符合高职学生认知规律，有助于实现有效教学。

　　本书打破传统的学科体系结构，将各知识点与操作技能恰当地融入各个项目中，突出现代职业教育的职业性和实践性，强化实践，培养学生实践动手能力，适应高职学生的学习特点，在教学过程中注意情感交流，因材施教，调动学生的学习积极性，提高教学效果。

　　本书是廊坊职业技术学院教师与企业工程师共同策划编写的一本工学结合教材。

　　本书由邵长文主编，李振松、史建政担任副主编，其中项目一～项目十六由邵长文编写，项目十七～项目二十由史建政编写，项目二十一和项目二十二由李振松编写，项目二十三和项目二十四由刘学普编写，项目二十五和项目二十六由白树成编写。全书由邵长文统稿。在本书的编写过程中，张昕教授提出了许多宝贵意见，电子工业出版社的王羽佳编辑为本书的出版做了大量工作。在此一并表示感谢！

　　由于计算机网络技术的迅猛发展和作者的水平有限，书中难免有错误和不妥之处，恳请读者批评指正。

编　者
2014 年 7 月

目　　录

第一篇　交换部分

项目一　MAC 地址表与地址端口绑定 ⋯⋯⋯⋯⋯⋯⋯⋯⋯⋯⋯⋯⋯⋯⋯⋯⋯⋯⋯⋯⋯⋯ 1
项目二　端口配置 ⋯⋯⋯⋯⋯⋯⋯⋯⋯⋯⋯⋯⋯⋯⋯⋯⋯⋯⋯⋯⋯⋯⋯⋯⋯⋯⋯⋯⋯⋯ 6
项目三　端口镜像 ⋯⋯⋯⋯⋯⋯⋯⋯⋯⋯⋯⋯⋯⋯⋯⋯⋯⋯⋯⋯⋯⋯⋯⋯⋯⋯⋯⋯⋯⋯ 16
项目四　端口汇聚 ⋯⋯⋯⋯⋯⋯⋯⋯⋯⋯⋯⋯⋯⋯⋯⋯⋯⋯⋯⋯⋯⋯⋯⋯⋯⋯⋯⋯⋯⋯ 24
项目五　VLAN 级联静态配置 ⋯⋯⋯⋯⋯⋯⋯⋯⋯⋯⋯⋯⋯⋯⋯⋯⋯⋯⋯⋯⋯⋯⋯⋯⋯ 30
项目六　VLAN 级联动态配置 ⋯⋯⋯⋯⋯⋯⋯⋯⋯⋯⋯⋯⋯⋯⋯⋯⋯⋯⋯⋯⋯⋯⋯⋯⋯ 34
项目七　VLAN 的 Hybrid 端口配置 ⋯⋯⋯⋯⋯⋯⋯⋯⋯⋯⋯⋯⋯⋯⋯⋯⋯⋯⋯⋯⋯⋯ 48
项目八　VLAN 间静态路由配置 ⋯⋯⋯⋯⋯⋯⋯⋯⋯⋯⋯⋯⋯⋯⋯⋯⋯⋯⋯⋯⋯⋯⋯⋯ 53
项目九　VLAN 间 RIP 协议配置 ⋯⋯⋯⋯⋯⋯⋯⋯⋯⋯⋯⋯⋯⋯⋯⋯⋯⋯⋯⋯⋯⋯⋯⋯ 58
项目十　高级 STP 配置 ⋯⋯⋯⋯⋯⋯⋯⋯⋯⋯⋯⋯⋯⋯⋯⋯⋯⋯⋯⋯⋯⋯⋯⋯⋯⋯⋯⋯ 64

第二篇　路由部分

项目十一　OSPF 基本配置 ⋯⋯⋯⋯⋯⋯⋯⋯⋯⋯⋯⋯⋯⋯⋯⋯⋯⋯⋯⋯⋯⋯⋯⋯⋯⋯ 78
项目十二　DR 的选举过程 ⋯⋯⋯⋯⋯⋯⋯⋯⋯⋯⋯⋯⋯⋯⋯⋯⋯⋯⋯⋯⋯⋯⋯⋯⋯⋯ 88
项目十三　虚链路、路由聚合和路由引入 ⋯⋯⋯⋯⋯⋯⋯⋯⋯⋯⋯⋯⋯⋯⋯⋯⋯⋯⋯ 93
项目十四　OSPF 的默认路由 ⋯⋯⋯⋯⋯⋯⋯⋯⋯⋯⋯⋯⋯⋯⋯⋯⋯⋯⋯⋯⋯⋯⋯⋯⋯ 104
项目十五　配置 OSPF 的 Stub 区域 ⋯⋯⋯⋯⋯⋯⋯⋯⋯⋯⋯⋯⋯⋯⋯⋯⋯⋯⋯⋯⋯⋯ 110
项目十六　配置 OSPF 的 NSSA 区域 ⋯⋯⋯⋯⋯⋯⋯⋯⋯⋯⋯⋯⋯⋯⋯⋯⋯⋯⋯⋯⋯ 114
项目十七　BGP 的基本配置 ⋯⋯⋯⋯⋯⋯⋯⋯⋯⋯⋯⋯⋯⋯⋯⋯⋯⋯⋯⋯⋯⋯⋯⋯⋯ 118
项目十八　BGP 的路由聚合 ⋯⋯⋯⋯⋯⋯⋯⋯⋯⋯⋯⋯⋯⋯⋯⋯⋯⋯⋯⋯⋯⋯⋯⋯⋯ 121
项目十九　BGP LOCAL-PREF 与 MED 属性的应用 ⋯⋯⋯⋯⋯⋯⋯⋯⋯⋯⋯⋯⋯⋯ 129
项目二十　BGP 路由反射 ⋯⋯⋯⋯⋯⋯⋯⋯⋯⋯⋯⋯⋯⋯⋯⋯⋯⋯⋯⋯⋯⋯⋯⋯⋯⋯ 136
项目二十一　基于 AS_PATH 的路由策略 ⋯⋯⋯⋯⋯⋯⋯⋯⋯⋯⋯⋯⋯⋯⋯⋯⋯⋯⋯ 139
项目二十二　基于 Community 属性的路由策略 ⋯⋯⋯⋯⋯⋯⋯⋯⋯⋯⋯⋯⋯⋯⋯⋯ 145
项目二十三　引入其他路由协议 ⋯⋯⋯⋯⋯⋯⋯⋯⋯⋯⋯⋯⋯⋯⋯⋯⋯⋯⋯⋯⋯⋯⋯ 149
项目二十四　OSPF 路由协议过滤接收的路由信息 ⋯⋯⋯⋯⋯⋯⋯⋯⋯⋯⋯⋯⋯⋯⋯ 154
项目二十五　RIP 过滤发布路由信息 ⋯⋯⋯⋯⋯⋯⋯⋯⋯⋯⋯⋯⋯⋯⋯⋯⋯⋯⋯⋯⋯ 159
项目二十六　VRRP 协议 ⋯⋯⋯⋯⋯⋯⋯⋯⋯⋯⋯⋯⋯⋯⋯⋯⋯⋯⋯⋯⋯⋯⋯⋯⋯⋯ 164
实训报告的基本内容及要求 ⋯⋯⋯⋯⋯⋯⋯⋯⋯⋯⋯⋯⋯⋯⋯⋯⋯⋯⋯⋯⋯⋯⋯⋯⋯⋯ 172

第一篇 交换部分

项目一 MAC地址表与地址端口绑定

1.1 项目提出

小王在某公司上班,发现员工随意更改计算机连接端口,或者拿未记录的设备连接公司网络,为了确保内网安全,想把所有员工的计算机和相应的端口进行一一对应,确保公司内网安全,他应该怎么办?

1.2 项目分析

1. 项目实训目的

- 掌握 E126 交换机中的 MAC 地址表的查看方法,了解 MAC 地址表中各表项的意义;
- 了解 MAC 地址的学习和老化过程;
- 了解 MAC 地址表的维护和管理方法:如何添加和删除表项;
- 掌握地址端口绑定的基本方法。

2. 项目实现功能

实现 MAC 地址与交换机端口绑定。

3. 项目主要应用的技术介绍

MAC 地址与交换机端口绑定其实就是交换机端口的安全功能。端口安全功能能让您配置一个端口只允许一台或者几台确定的设备访问那个交换机;能根据 MAC 地址确定允许访问的设备;允许访问的设备的 MAC 地址既可以手工配置,也可以从交换机"学到";当一个未批准的 MAC 地址试图访问端口的时候,交换机会挂起或者禁用该端口等。

4. 首先必须明白两个概念

静态可靠的 MAC 地址:在交换机接口模式下手动配置,这个配置会被保存在交换机 MAC 地址表和运行配置文件中,交换机重新启动后不丢失(当然是在保存配置完成后)。

动态可靠的 MAC 地址:这种类型是交换机默认的类型。在这种类型下,交换机会动态学习 MAC 地址,但是这个配置只会保存在 MAC 地址表中,不会保存在运行配置文件中,并且交换机重新启动后,这些 MAC 地址表中的 MAC 地址自动会被清除。

1.3 项目实施

1. 项目拓扑图

MAC 地址表与地址端口绑定如图 1-1 所示。

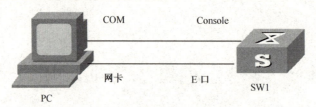

图 1-1　MAC 地址表与地址端口绑定

2. 项目实训环境准备

E126（1 台）、计算机（1 台）。

为了不受原来的配置影响，在实训之前先将所有的配置数据擦除后重新启动，命令为：

<h3c>Reset saved-configuration
<h3c>reboot

3. 项目主要实训步骤

（1）查看 MAC 地址表。

使用 display mac-address 命令，可以查看连接在交换机上的 PC1 的 MAC 地址：

```
[sw1]display mac-address
MAC ADDR          VLAN ID     STATE        PORT INDEX        AGING TIME(s)
001e-9001-89b1       1        Learned      Ethernet1/0/4     AGING
   ---  1 mac address(es) found  ---
```

MAC ADDR：表示 Ethernet1/0/4 所连接的 PC1 的 MAC 地址。

VLAN ID：表示这个端口所在的 VLAN。

STATE：表示这个 MAC 地址表项的属性，Learned 表示 MAC 地址为动态学习的。

PORT INDEX：这是物理端口号。

AGING TIME(s)：表示该表项的老化时间，这个表项还会被保存多长时间（以秒为单位）。

交换机中正是由于 MAC 地址表的存在，才使得交换机可以根据数据包的 MAC 地址查出端口号，来实现基于二层的快速转发。

当一台 PC 被连接到交换机的端口，交换机会从该端口动态学习到 PC 的 MAC 地址，并将其添加到 MAC 地址表中。同时，可以在交换机的 ARP 表中看到这台 PC 的 IP 和 MAC。

（2）查看 ARP 表。

①首先，先断开 PC1 和 SW1 的网线，设置 Vlan 1 的 IP 地址，192.168.1.200/24，如果没有配置，查看不到 ARP 表的信息。

```
[sw1]interface Vlan-interface 1
[sw1-Vlan-interface1]ip address 192.168.1.200 24
[sw1-Vlan-interface1]quit
```

②使用 debugging arp packet 和 terminal debugging 命令打开 ARP 的学习过程。然后连接 PC1 到 SW1 上，可以看到很多交换的信息包，从中可以观察学习 MAC 得知的过程。

```
<sw1>debugging arp packet
<sw1>terminal debugging
% Current terminal debugging is on
*0.26286199 sw1 ARP/8/arp_rcv:- 1 -Receive an ARP Packet, operation : 1,
sender_0-0000-0000,
target_ip_addr : 192.168.1.129
*0.26286435 sw1 ARP/8/arp_status_change:- 1 -ARP Item Status Change :
eth_addr : 001e-9001-89b1,
ip_addr : 192.168.1.129,
INITIALIZE -> NO_AGE
```

使用 undo terminal debugging 可以关闭调试。

```
<sw1>undo terminal debugging
% Current terminal debugging is off
```

使用 display arp 命令可以查看交换机的 ARP 表的内容。从下表可以看到，PC1 的 IP 和 MAC 地址等信息：

```
<sw1>display arp
          Type: S-Static    D-Dynamic
IP Address       MAC Address      VLAN ID    Port Name / AL ID    Aging    Type
192.168.1.129    001e-9001-89b1   1          Ethernet1/0/4        18       D
---    1 entry found    ---
```

以上显示 PC1 的 IP 地址，MAC 地址，端口所属 VLAN，对应端口，老化时间，表项类型。其中表中的 TPYE 有两种：一种是 D-Dynamic，表示动态学习的；另一种是 S-Static，表示静态配置的。而表中老化时间 AGING 表示这个表项还会被保存多长时间。系统默认的老化时间是 300 秒。可以通过下列命令来查看 MAC 地址的老化时间：

```
<sw1>display mac-address aging-time
Mac address aging time: 300s
```

同时还可以在系统视图下通过 mac-address timer aging 命令来修改系统的老化时间。默认情况下，学习到的地址表项在老化时间结束后会被删除。当断开 PC 与交换机的物理连接时，这个动态表项会立即删除（不管老化时间是否已经到了）。

```
[sw1]mac-address timer aging ?
    INTEGER<10-1000000>    Global aging time (second)
[sw1]mac-address timer aging 200          //设置老化时间为 200 秒
```

（3）MAC 地址表的维护和管理。

交换机除了可以动态从端口动态学习到 MAC 地址之外，还可以通过命令手动添加和删除静态表项，来实现对 MAC 地址表的维护和控制。在 E126 交换机中，对于同一个 MAC 地址只能有一个表项，因此在配置静态表项之前，必须把相应的动态表项释放掉，然后再进行如下配置：

```
[sw1]mac-address static 001e-9001-89b1 interface Ethernet 1/0/4 vlan 1
[sw1]display mac-address
MAC ADDR          VLAN ID      STATE          PORT INDEX        AGING TIME(s)
001e-9001-89b1       1        Config static   Ethernet1/0/4      NOAGED
```

从上面的 MAC 地址表中可以看出，STATE 字段是 Config static，表示是手动添加的静态表项，AGING TIME(s)为 NOAGED，表示没有老化时间。只要交换机没有重启丢失配置，该表项就会一直存在。

可以用"undo"命令删除配置的静态表项，配置如下：

```
[sw1]undo mac-address static 001e-9001-89b1 interface Ethernet 1/0/4 vlan 1
[sw1]dis mac-address
No MAC addresses found.
```

除了可以配置静态表项之外，还可以手动配置动态表项，配置命令如下：

```
[sw1]mac-address dynamic 001e-9001-89b1 interface Ethernet1/0/4 vlan 1
[sw1]display mac- address
MAC ADDR          VLAN ID      STATE          PORT INDEX        AGING TIME(s)
001e-9001-89b1       1        Config dynamic  Ethernet1/0/4      AGING
```

请注意：这里的 STATE 是 Config dynamic 表示是手动添加的动态表项。手动添加的动态表项和自动学习的动态表项一样有老化时间。同样也可以用"undo"命令来删除动态表项：

```
[sw1]undo mac-address dynamic 001e-9001-89b1 interface Ethernet1/0/4 vlan 1
[sw1]display mac- address
No MAC addresses found.
```

在交换机的接口配置视图下，可以用"mac-address max-mac-count"命令来配置端口可以学习的最大 MAC 地址数，配置如下：

```
[sw1-Ethernet1/0/4]mac-address max-mac-count 600        // E1/0/4 最多能学 600 个 MAC
```

默认情况，端口最多可以学习 4096 个 MAC。

（4）地址端口绑定。

在交换机上，有两种方法可以实现地址端口绑定。在做实训之前，请将另外一台 PC2 连接到交换机上，端口是 E1/0/2，IP 是 192.168.1.130/24。

为了把 PC1 和端口 E1/0/4 捆绑起来，首先配置一个静态 MAC 地址表项与 PC1 的 MAC 地址和 E1/0/4 对应，这样即使将 PC1 接到其他端口，由于静态配置优先于动态学习，同一个 MAC 地址只能有一个表项的原则，PC1 的 MAC 地址不能被其他端口学习到，也就是说 PC1 不能连接在其他端口上，否则不能通信。

```
[H3C]dis mac-ad
MAC ADDR          VLAN ID      STATE          PORT INDEX        AGING TIME(s)
001e-9004-0216       1        Learned        Ethernet1/0/2      AGING
001e-9001-89b1       1        Learned        Ethernet1/0/4      AGING
[H3C]mac-address static 001e-9001-89b1 interface Ethernet 1/0/4 vlan 1
[H3C]dis mac-ad
MAC ADDR          VLAN ID      STATE          PORT INDEX        AGING TIME(s)
```

```
001e-9004-0216    1    Learned          Ethernet1/0/2    AGING
001e-9001-89b1    1    Config static    Ethernet1/0/4    NOAGED
```

然后在 PC2 上 ping PC1，看是否能通：

C:\>ping 192.168.1.129
Pinging 192.168.1.129 with 32 bytes of data:
Reply from 192.168.1.129: bytes=32 time<1ms TTL=128
Reply from 192.168.1.129: bytes=32 time<1ms TTL=128
Reply from 192.168.1.129: bytes=32 time<1ms TTL=128
Reply from 192.168.1.129: bytes=32 time<1ms TTL=128
Ping statistics for 192.168.1.129:
    Packets: Sent = 4, Received = 4, Lost = 0 (0% loss),
Approximate round trip times in milli-seconds:
    Minimum = 0ms, Maximum = 0ms, Average = 0ms

接下来，将 PC1 连接到 E1/0/6，再次在 PC2 上 ping PC1，查看结果：

C:\>ping 192.168.1.129
Pinging 192.168.1.129 with 32 bytes of data:
Request timed out.
Request timed out.
Request timed out.
Request timed out.
Ping statistics for 192.168.1.129:
    Packets: Sent = 4, Received = 0, Lost = 4 (100% loss),

这时 E1/0/4 还有可能学习到其他的 MAC 地址，也就是说 E1/0/4 还可以连接其他的主机，为了让 E1/0/4 只能连接 PC1，需要在 E1/0/4 没有学习到其他 MAC 地址的时候，禁止这个端口进行地址学习。配置命令如下：

[H3C]int Ethernet 1/0/4
[H3C-Ethernet1/0/4]mac-address max-mac-count 0

这样 E1/0/4 就不会再学习其他地址。我们可以另外连接一台 PC3 到 E1/0/4 上来进行验证。请同学们考虑为什么要禁止学习，为什么一个端口没禁用之前还能学习其他地址？

1.4 项目总结与提高

（1）写出主要项目实施规划、步骤与实训所得的主要结论。
（2）搜索有关地址绑定，查看能否做到其他绑定功能，比如：PORT+IP、 PORT+IP+MAC、IP+MAC 等。

项目二 端口配置

2.1 项目提出

小王在某公司上班,发现有的计算机连接速度是 100Mbps,有的计算机是 10Mbps,这让他很困惑,这是怎么回事,他应该怎么办?

2.2 项目分析

1. 项目实训目的

理解并测试端口速率、工作方式自协商功能。

2. 项目实现功能

学会查看端口工作状态,理解端口速率、工作方式等功能。

3. 项目主要应用的技术介绍

(1) 交换机端口基础。

随着网络技术的不断发展,需要网络互联处理的事务越来越多,为了适应网络需求,以太网技术也完成了一代又一代的技术更新。为了兼容不同的网络标准,端口技术变得尤为重要。端口技术主要包含了端口自协商、网络智能识别、流量控制、端口聚合以及端口镜像等技术,它们很好地解决了各种以太网标准互连互通存在的问题。以太网主要有以下标准:标准以太网、快速以太网和千兆以太网等。它们分别有不同的端口速度和工作模式。

(2) 端口速率自协商。

标准以太网其端口速率为固定 10M。快速以太网支持的端口速率有 10M、100M 和自适应三种方式。千兆以太网支持的端口速率有 10M、100M、1000M 和自适应方式。以太网交换机支持端口速率的手工配置和自适应。默认情况下,所有端口都是自适应工作方式,通过相互交换自协商报文进行匹配。

当链路两端一端为自协商,另一端为固定速率时,我们建议修改两端的端口速率,保持端口速率一致。

如果两端都以固定的速率工作,而工作速率不一致时,很容易出现通信故障,这种现象应该尽量避免。

(3) 端口工作模式。

交换机端口有半双工和全双工两种端口模式。目前交换机可以手工配置,也可以自动协商来决定端口究竟工作在何种模式。

(4) 端口的接口类型。

目前以太网接口有 MDI 和 MDIX 两种类型。MDI 称为介质相关接口,MDIX 称为介质非相关接口。我们常见的以太网交换机所提供的端口都属于 MDIX 接口,而路由器和 PC 提供的都属于 MDI 接口。有的交换机同时支持上述两种接口，我们可以强制制定交换机端口的接口类型，其配置命令如下：

[h3c-Ethernet0/1] mdi {normal| cross| auto}
Normal：表示端口为 MDIX 接口
Cross：表示端口为 MDI 接口
Auto：表示端口工作在自协商模式

（5）流量控制。

由于标准以太网、快速以太网和千兆以太网混合组网，在某些网络接口不可避免会出现流量过大的现象而产生端口阻塞。为了减轻和避免端口阻塞的产生，标准协议专门规定了解决这一问题的流量控制技术。在交换机中所有端口缺省情况下都禁用了流量控制功能。开启/关闭流量控制功能的配置命令如下：

[h3c-Ethernet0/1]flow-control
[h3c-Ethernet0/1]undo flow-control

2.3 项目实施

1．项目拓扑图

端口配置如图 2-1 所示。

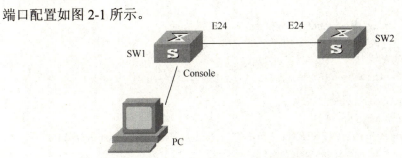

图 2-1　端口配置

2．项目实训环境准备

E126（2 台）、计算机（1 台）。

为了不受原来的配置影响，在实训之前先将所有的配置数据擦除后重新启动，命令为：

<h3c>Reset saved-configuration
<h3c>reboot

3．项目主要实训步骤

按照拓扑图连接所有设备。检查设备的软件版本及配置信息，所有配置为初始状态。如果配置不符合要求，请在用户视图下擦除设备中的配置文件（reset saved-configuration），然后重启设备（reboot）以使系统采用缺省的配置参数进行初始化。

（1）初始配置。

为了测试两台交换机的连通性，首先给每台交换机配置一个 3 层接口，并分配 IP 地址 192.168.1.1 /24 和 192.168.1.2 /24。然后通过 PING 命令测试互通性。

SW1 的配置：

```
[H3C]sys sw1
[sw1]int Vlan-interface 1
[sw1-Vlan-interface1]ip add 192.168.1.1 24
```

SW2 的配置：

```
[H3C]sys sw2
[sw2]int v 1
[sw2-Vlan-interface1]ip add 192.168.1.2 24
```

配置完成后，在 SW1 上 ping SW2，检查互通性：

```
[sw1]ping 192.168.1.2
  PING 192.168.1.2: 56    data bytes, press CTRL_C to break
    Reply from 192.168.1.2: bytes=56 Sequence=1 ttl=255 time=671 ms
    Reply from 192.168.1.2: bytes=56 Sequence=2 ttl=255 time=302 ms
    Reply from 192.168.1.2: bytes=56 Sequence=3 ttl=255 time=319 ms
    Reply from 192.168.1.2: bytes=56 Sequence=4 ttl=255 time=305 ms
    Reply from 192.168.1.2: bytes=56 Sequence=5 ttl=255 time=314 ms
  --- 192.168.1.2 ping statistics ---
    5 packet(s) transmitted
    5 packet(s) received
    0.00% packet loss
    round-trip min/avg/max = 302/382/671 ms
```

（2）端口速率的自适应特性。

用命令可以查看到当前端口的速率状况，如下所示为 SW1 的 E1/0/24 的速率状况：

```
[sw1]display interface Ethernet 1/0/24
  Ethernet1/0/24 current state : UP
  IP Sending Frames' Format is PKTFMT_ETHNT_2, Hardware address is 000f-e25c-40f6
  Media type is twisted pair, loopback not set
  Port hardware type is 100_BASE_TX
  **100Mbps-speed mode**, full-duplex mode
  **Link speed type is autonegotiation**, link duplex type is autonegotiation
  Flow-control is not enabled
  The Maximum Frame Length is 1536
  Broadcast MAX-ratio: 100%
  PVID: 1
  Mdi type: normal
  Port link-type: access
   Tagged    VLAN ID : none
   Untagged VLAN ID : 1
```

上面输出部分指出了此端口工作于速率自动协商方式下，协商速率为 100Mbps。因为 SW1 和 SW2 的 E1/0/24 都是 100M/10M 自适应端口。默认情况下它们都工作于速率自协商方式下，所以结果是它们支持的最高速率：100Mbps。

可以用命令改变端口速率工作方式，例如改变 SW1 的端口速率为 10Mbps：

```
[sw1]int e1/0/24
```

[sw1-Ethernet1/0/24]speed 10

[sw1-Ethernet1/0/24]

#Apr　　2 00:35:10:315 2000 sw1 L2INF/2/PORT LINK STATUS CHANGE:- 1 - Trap 1.3.6.1.6.3.1.1.5.3: portIndex is 4227810, ifAdminStatus is 1, ifOperStatus is 2

%Apr　　2 00:35:10:492 2000 sw1 L2INF/5/PORT LINK STATUS CHANGE:- 1 - Ethernet1/0/24: is **DOWN**

[sw1-Ethernet1/0/24]

#Apr　　2 00:35:13:013 2000 sw1 L2INF/2/PORT LINK STATUS CHANGE:- 1 - Trap 1.3.6.1.6.3.1.1.5.4: portIndex is 4227810, ifAdminStatus is 1, ifOperStatus is 1

%Apr　　2 00:35:13:192 2000 sw1 L2INF/5/PORT LINK STATUS CHANGE:- 1 - Ethernet1/0/24: is **UP**

执行完这条命令后，测试 SW1 和 SW2 的连通性：

[sw1]ping 192.168.1.2

　　PING 192.168.1.2: 56　　data bytes, press CTRL_C to break

　　　　Reply from 192.168.1.2: bytes=56 Sequence=1 ttl=255 time=624 ms

　　　　Reply from 192.168.1.2: bytes=56 Sequence=2 ttl=255 time=321 ms

　　　　Reply from 192.168.1.2: bytes=56 Sequence=3 ttl=255 time=313 ms

　　　　Reply from 192.168.1.2: bytes=56 Sequence=4 ttl=255 time=303 ms

　　　　Reply from 192.168.1.2: bytes=56 Sequence=5 ttl=255 time=324 ms

　　--- 192.168.1.2 ping statistics ---

　　　　5 packet(s) transmitted

　　　　5 packet(s) received

　　　　0.00% packet loss

　　　　round-trip min/avg/max = 303/377/624 ms

在 SW1 上，display interface Ethernet 1/0/24 如下所示，24 口端口速率是 10Mbps，类型是 force link。

[sw1]display interface Ethernet 1/0/24

　　Ethernet1/0/24 current state : UP

　　IP Sending Frames' Format is PKTFMT_ETHNT_2, Hardware address is 000f-e25c-40f6

　　Media type is twisted pair, loopback not set

　　Port hardware type is 100_BASE_TX

　　10Mbps-speed mode, full-duplex mode

　　Link speed type is force link, link duplex type is autonegotiation

　　Flow-control is not enabled

　　The Maximum Frame Length is 1536

　　Broadcast MAX-ratio: 100%

　　PVID: 1

　　Mdi type: normal

　　Port link-type: access

　　　Tagged　　　VLAN ID : none

　　　Untagged VLAN ID : 1

在 SW2 上，display interface Ethernet 1/0/24 如下所示，24 口端口速率是 10Mbps，类型是 autonegotiation。

[sw2]display interface Ethernet 1/0/24

　　Ethernet1/0/24 current state : UP

```
Ethernet1/0/24 current state : UP
Media type is twisted pair, loopback not set
Port hardware type is 100_BASE_TX
```
10Mbps-speed mode, full-duplex mode
Link speed type is autonegotiation, link duplex type is autonegotiation
```
Flow-control is not enabled
The Maximum Frame Length is 9216
Broadcast MAX-ratio: 100%
Unicast MAX-ratio: 100%
Multicast MAX-ratio: 100%
Allow jumbo frame to pass
PVID: 1
Mdi type: auto
Port link-type: access
  Tagged    VLAN ID : none
  Untagged VLAN ID : 1
```

也就是说，现在 SW1 和 SW2 工作在 10Mbps 速率下，如果我们强制 SW2 的端口速率为 100Mbps，结果会如何？

```
[sw2]int e1/0/24
[sw2-Ethernet1/0/24]speed 100
```

SW2 ping SW1 结果如下：

```
[sw2-Ethernet1/0/24]ping 192.168.1.1
  PING 192.168.1.1: 56   data bytes, press CTRL_C to break
    Request time out
    Request time out
    Request time out
    Request time out
    Request time out
  --- 192.168.1.1 ping statistics ---
    5 packet(s) transmitted
    0 packet(s) received
    100.00% packet loss

[sw2-Ethernet1/0/24]dis int e1/0/24
  Ethernet1/0/24 current state :
```
DOWN
```
Media type is twisted pair, loopback not set
 Port hardware type is 100_BASE_TX
 100Mbps-speed mode, unknown-duplex mode
 Link speed type is force link, link duplex type is autonegotiation
 Flow-control is not enabled
 The Maximum Frame Length is 9216
 Broadcast MAX-ratio: 100%
 Unicast MAX-ratio: 100%
 Multicast MAX-ratio: 100%
 Allow jumbo frame to pass
```

PVID: 1
   Mdi type: auto
   Port link-type: access
    Tagged   VLAN ID : none
    Untagged VLAN ID : 1

如上显示，SW2 的 24 口是 DOWN 的，也就是说当两台交换机的两个端口工作于不一样的速率下，端口是 DOWN 的。这个现象在实际网络故障中是比较常见的，请一定注意。

进入下一步骤前，先恢复交换机的端口状态，接口视图下用 undo speed 命令：

   [sw1-Ethernet1/0/24]undo speed
   [sw2-Ethernet1/0/24]undo speed

并使用 ping 命令测试连通性。

   [sw1]ping 192.168.1.2
     PING 192.168.1.2: 56   data bytes, press CTRL_C to break
       Reply from 192.168.1.2: bytes=56 Sequence=1 ttl=255 time=663 ms
       Reply from 192.168.1.2: bytes=56 Sequence=2 ttl=255 time=320 ms
       Reply from 192.168.1.2: bytes=56 Sequence=3 ttl=255 time=332 ms
       Reply from 192.168.1.2: bytes=56 Sequence=4 ttl=255 time=323 ms
       Reply from 192.168.1.2: bytes=56 Sequence=5 ttl=255 time=318 ms
   --- 192.168.1.2 ping statistics ---
     5 packet(s) transmitted
     5 packet(s) received
     0.00% packet loss
     round-trip min/avg/max = 318/391/663 ms

（3）端口工作方式的自适应特性。

现在来测试交换机的工作方式，它和速率协商特性类似。交换机的端口工作方式可以是自动协商模式、全双工模式和半双工模式，默认是自动协商模式。

先查看 SW1 的 24 端口工作方式：

   [sw1]display interface Ethernet 1/0/24
     Ethernet1/0/24 current state : UP
     IP Sending Frames' Format is PKTFMT_ETHNT_2, Hardware address is 000f-e25c-40f6
     Media type is twisted pair, loopback not set
     Port hardware type is 100_BASE_TX
     100Mbps-speed mode, **full-duplex mode**
     Link speed type is autonegotiation, **link duplex type is autonegotiation**
     Flow-control is not enabled
     The Maximum Frame Length is 1536
     Broadcast MAX-ratio: 100%
     PVID: 1
     Mdi type: normal
     Port link-type: access
      Tagged   VLAN ID : none
      Untagged VLAN ID : 1

如上所示，当前工作方式为自动协商模式，协商所得为全双工 full-duplex。可以用 duplex 命令配置当前接口的工作方式：

    [sw1]int e1/0/24
    [sw1-Ethernet1/0/24]duplex half

SW1 的 24 口状态：

    [sw1-Ethernet1/0/24]dis int e1/0/24
      Ethernet1/0/24 current state : UP
      IP Sending Frames' Format is PKTFMT_ETHNT_2, Hardware address is 000f-e25c-40f6
      Media type is twisted pair, loopback not set
      Port hardware type is 100_BASE_TX
      100Mbps-speed mode, **half-duplex mode**
      Link speed type is autonegotiation, **link duplex type is force link**
      Flow-control is not enabled
      The Maximum Frame Length is 1536
      Broadcast MAX-ratio: 100%
      PVID: 1
      Mdi type: normal
      Port link-type: access
        Tagged   VLAN ID : none
        Untagged VLAN ID : 1

SW2 的 24 口状态：

      Ethernet1/0/24 current state : UP
      IP Sending Frames' Format is PKTFMT_ETHNT_2, Hardware address is 000f-e254-8708
      Media type is twisted pair, loopback not set
      Port hardware type is 100_BASE_TX
      100Mbps-speed mode, **half-duplex mode**
      Link speed type is autonegotiation, **link duplex type is autonegotiation**
      Flow-control is not enabled
      The Maximum Frame Length is 9216
      Broadcast MAX-ratio: 100%
      Unicast MAX-ratio: 100%
      Multicast MAX-ratio: 100%
      Allow jumbo frame to pass
      PVID: 1
      Mdi type: auto
      Port link-type: access
        Tagged   VLAN ID : none
        Untagged VLAN ID : 1

测试连通性：

    [sw1]ping 192.168.1.2
      PING 192.168.1.2: 56   data bytes, press CTRL_C to break
        Reply from 192.168.1.2: bytes=56 Sequence=1 ttl=255 time=370 ms
        Reply from 192.168.1.2: bytes=56 Sequence=2 ttl=255 time=370 ms

Reply from 192.168.1.2: bytes=56 Sequence=3 ttl=255 time=371 ms
Reply from 192.168.1.2: bytes=56 Sequence=4 ttl=255 time=369 ms
Reply from 192.168.1.2: bytes=56 Sequence=5 ttl=255 time=359 ms
--- 192.168.1.2 ping statistics ---
5 packet(s) transmitted
5 packet(s) received
0.00% packet loss
round-trip min/avg/max = 359/367/371 ms

如上所示，SW1 上 24 口的 half-duplex mode 是手工配置的，SW2 上 24 口的 half-duplex mode 是自动协商的，连通性正常。

如果把 SW2 的 24 口设置工作于全双工，结果会怎样？

[sw2-Ethernet1/0/24]duplex full
[sw2]dis int e1/0/24
Ethernet1/0/24 current state : UP
IP Sending Frames' Format is PKTFMT_ETHNT_2, Hardware address is 000f-e254-8708
Media type is twisted pair, loopback not set
Port hardware type is 100_BASE_TX
100Mbps-speed mode, **full-duplex mode**
Link speed type is autonegotiation, **link duplex type is force link**
Flow-control is not enabled
The Maximum Frame Length is 9216
Broadcast MAX-ratio: 100%
Unicast MAX-ratio: 100%
Multicast MAX-ratio: 100%
Allow jumbo frame to pass
PVID: 1
Mdi type: auto
Port link-type: access
 Tagged   VLAN ID : none
 Untagged VLAN ID : 1

测试连通性，SW2 ping SW1：

[sw2]ping 192.168.1.1
PING 192.168.1.1: 56   data bytes, press CTRL_C to break
Reply from 192.168.1.1: bytes=56 Sequence=1 ttl=254 time=757 ms
Reply from 192.168.1.1: bytes=56 Sequence=2 ttl=254 time=373 ms
Reply from 192.168.1.1: bytes=56 Sequence=3 ttl=254 time=374 ms
Reply from 192.168.1.1: bytes=56 Sequence=4 ttl=254 time=366 ms
Reply from 192.168.1.1: bytes=56 Sequence=5 ttl=254 time=370 ms
--- 192.168.1.1 ping statistics ---
5 packet(s) transmitted
5 packet(s) received
0.00% packet loss
round-trip min/avg/max = 366/448/757 ms

结果显示：两台交换机是连通的。不要被这个现象蒙骗，如果在 SW1 和 SW2 上使用大包、连续多次 ping 对方，就会出现丢包现象：

    [sw1]ping -s 1000 -c 1000 192.168.1.2
      PING 192.168.1.2: 1000   data bytes, press CTRL_C to break
        Reply from 192.168.1.2: bytes=1000 Sequence=1 ttl=255 time=374 ms
        Reply from 192.168.1.2: bytes=1000 Sequence=2 ttl=255 time=371 ms
        Reply from 192.168.1.2: bytes=1000 Sequence=3 ttl=255 time=371 ms
        Reply from 192.168.1.2: bytes=1000 Sequence=4 ttl=255 time=375 ms
        …
        Reply from 192.168.1.2: bytes=1000 Sequence=6 ttl=255 time=382 ms
        Request time out
        Reply from 192.168.1.2: bytes=1000 Sequence=8 ttl=255 time=374 ms
        Reply from 192.168.1.2: bytes=1000 Sequence=9 ttl=255 time=370 ms
      …
      --- 192.168.1.2 ping statistics ---
        1000 packet(s) transmitted
        999 packet(s) received
        0.10% packet loss
        round-trip min/avg/max = 4/148/390 ms

如上所示，丢包现象比较严重。在实际的网络中，双向数据流是很常见的需求，所以必须注意上面案例所示的这类比较隐蔽的故障。同样，用如下命令清除这个实训步骤对于下面步骤的影响：

    [sw1-Ethernet1/0/24]undo duplex
    [sw2-Ethernet1/0/24]undo duplex

然后，用 ping 命令测试 SW1 和 SW2 的连通性。

    [sw1]ping -s 1000 -c 1000 192.168.1.2
    …
    --- 192.168.1.2 ping statistics ---
      1000 packet(s) transmitted
      1000 packet(s) received
      0.00% packet loss
      round-trip min/avg/max = 4/148/390 ms

【参考资料】

    [sw2]ping ?

选项	说明
-a	指定 ping 命令的源地址
-c	设置发送 ICMP ECHO_REQUEST 报文的数目
-d	打开所使用套接口上的 SO_DEBUG 选项
-f	指定发送数据包不能被分片
-h	指定发送回显请求报文的 TTL 值
-i	指定 ping 命令的发送接口
-n	不对目的主机作域名解析

-p	设置对于发送的 ECHO_REQUEST 报文的填充字节，长度不超过 8 个 16 进制字符，例如, -p f2 将报文全部填充为 f2
-q	除统计数字外，不显示其他的详细信息
-s	设置 ECHO_REQUEST 报文的长度
-t	设置等待 ECHO_REQUEST 报文响应的超时时间
-tos	指定发送回显请求报文的 TOS 值
-v	显示接收到的非 ECHO_RESPONSE ICMP 报文，缺省是不显示
STRING<1-30>	远程系统的 IP 地址或主机名
ip	IP 协议

2.4 项目总结与提高

（1）写出主要项目实施规划、步骤与实训所得的主要结论。
（2）请课余时间详细研究流量控制是如何实现的。

项目三 端口镜像

3.1 项目提出

小王在某公司上班,出于信息安全、保护公司机密的需要,想了解网络流量的变化规律,监视到进出网络的所有数据包,迫切需要网络中有一个端口能提供这种实时监控功能。他应该怎么办?

3.2 项目分析

1. 项目实训目的

掌握端口镜像功能。

2. 项目实现功能

利用端口镜像功能进行端口通信观察。

3. 项目主要应用的技术介绍

端口镜像(Port Mirroring)把交换机一个或多个端口(VLAN)的数据镜像到一个或多个端口的方法。端口镜像又称端口映射,是网络通信协议的一种方式。

3.3 项目实施

1. 项目拓扑图

端口镜像如图 3-1 所示。

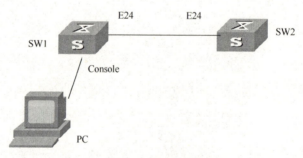

图 3-1 端口镜像

2. 项目实训环境准备

E126(2 台)、计算机(1 台)。

为了不受原来的配置影响,在实训之前先将所有的配置数据擦除后重新启动,命令为:

    <h3c>Reset saved-configuration
    <h3c>reboot

3. 项目主要实训步骤

按照拓扑图连接所有设备。检查设备的软件版本及配置信息,所有配置为初始状态。如果配置不符合要求,请在用户视图下擦除设备中的配置文件(reset saved-configuration),然后重启设备(reboot)以使系统采用缺省的配置参数进行初始化。

(1) 初始配置。

为了测试两台交换机的连通性,首先给每台交换机配置一个 3 层接口,并分配 IP 地址 192.168.1.1 /24 和 192.168.1.2 /24。然后通过 ping 命令测试互通性。

SW1 的配置:

```
[H3C]sys sw1
[sw1]int v
[sw1]int Vlan-interface 1
[sw1-Vlan-interface1]ip add 192.168.1.1 24
```

SW2 的配置:

```
[H3C]sys sw2
[sw2]int v 1
[sw2-Vlan-interface1]ip add 192.168.1.2 24
```

配置完成后,在 SW1 上 ping SW2,检查互通性:

```
[sw1]ping 192.168.1.2
  PING 192.168.1.2: 56   data bytes, press CTRL_C to break
    Reply from 192.168.1.2: bytes=56 Sequence=1 ttl=255 time=671 ms
    Reply from 192.168.1.2: bytes=56 Sequence=2 ttl=255 time=302 ms
    Reply from 192.168.1.2: bytes=56 Sequence=3 ttl=255 time=319 ms
    Reply from 192.168.1.2: bytes=56 Sequence=4 ttl=255 time=305 ms
    Reply from 192.168.1.2: bytes=56 Sequence=5 ttl=255 time=314 ms
  --- 192.168.1.2 ping statistics ---
    5 packet(s) transmitted
    5 packet(s) received
    0.00% packet loss
    round-trip min/avg/max = 302/382/671 ms
```

(2) 端口镜像。

镜像一般是将符合指定规则的报文复制到镜像目的端口。一般镜像目的端口会接入数据检测设备,用户利用这些设备对镜像过来的报文进行分析,进行网络监控和故障排除等。

流镜像就是将匹配 ACL 规则的业务流复制到指定的目的端口,用于报文分析和监视。在配置流镜像前用户需要先定义符合需求的 ACL 规则,设备会引用这些 ACL 规则进行流识别。

端口镜像,即将指定端口接收或者发送的报文复制到镜像目的端口。

确定了镜像源端口,再确定被镜像报文的方向:inbound 表示仅对端口接收的报文进行镜像,outbound 表示仅对端口发送的报文进行镜像,both 表示同时对端口接收和发送的报文进行镜像。

(3) E328 配置命令介绍。

E352&E328 交换机支持的镜像功能及相关命令见表 3-1。

表 3-1 E352&E328 交换机支持的镜像功能及相关命令

功能	规格	相关命令	
镜像	支持流镜像	monitor-port mirrored-to	
	支持端口镜像	monitor-port mirroring-port	

①配置流镜像。

配置准备：

- 定义了进行流识别的 ACL。关于定义 ACL 的描述请参见本书"ACL"模块的描述
- 确定了镜像目的端口
- 确定需要进行流镜像配置的端口和被镜像流的方向

配置过程见表 3-2。

表 3-2 配置流镜像

操作	命令	说明
进入系统视图	system-view	-
进入镜像目的端口的以太网端口视图	interface *interface-type interface-number*	-
定义当前端口为镜像目的端口	monitor-port	必选 镜像目的端口上不能使能 LACP 及 STP
退出当前视图	quit	-
进入进行流镜像配置的以太网端口视图	interface *interface-type interface-number*	-
引用 ACL 进行流识别，对匹配的报文进行流镜像	mirrored-to { inbound | outbound } *acl-rule* { monitor-interface | cpu }	必选
显示流量镜像的参数设置	display qos-interface { *interface-type interface-number* | *unit-id* } mirrored-to	可选 display 命令可以在任意视图下执行
显示端口的所有 QoS 设置信息	display qos-interface { *interface-type interface-number* | *unit-id* } all	

acl-rule：下发的 ACL，可以是多种 ACL 的组合。组合方式说明见表 3-3。

表 3-3 组合下发 ACL 的方式

组合方式	acl-rule 的形式
单独下发一个 IP 型 ACL(包括基本 ACL 与高级 ACL)中所有子规则	ip-group *acl-number*
单独下发一个 IP 型 ACL 中一条子规则	ip-group *acl-number* rule *rule-id*
单独下发一个二层 ACL 中所有子规则	link-group *acl-number*
单独下发一个二层 ACL 中一条子规则	link-group *acl-number* rule *rule-id*
单独下发一个用户自定义 ACL 中所有子规则	user-group *acl-number*
单独下发一个用户自定义 ACL 中一条子规则	user-group *acl-number* rule *rule-id*
同时下发 IP 型 ACL 中一条子规则和二层型 ACL 的一条子规则	ip-group *acl-number* rule *rule-id* link-group *acl-number* rule *rule-id*

②配置端口镜像。

配置准备：

- 确定了镜像源端口，确定了被镜像报文的方向：inbound 表示仅对端口接收的报文进

行镜像，outbound 表示仅对端口发送的报文进行镜像，both 表示同时对端口接收和发送的报文进行镜像
- 确定了镜像目的端口

配置过程见表 3-4。

表 3-4　配置端口镜像

操作	命令	说明
进入系统视图	system-view	-
进入镜像目的端口的以太网端口视图	interface *interface-type interface-number*	-
定义当前端口为镜像目的端口	monitor-port	必选 镜像目的端口上不能使能 LACP 及 STP
退出当前视图	quit	-
进入镜像源端口的以太网端口视图	interface *interface-type interface-number*	-
配置镜像源端口，同时指定被镜像报文的方向	mirroring-port { inbound | outbound | both }	必选
显示镜像的参数设置	display mirror	可选 display 命令可以在任意视图下执行

（4）E126 配置命令介绍。

E126 以太网交换机业务特性：支持一对多的端口镜像（即一个镜像端口，而对被镜像端口的数量没有限制），支持 RSPAN（Remote Switched Port Analyzer，远程交换端口分析）。

①在以太网端口视图下配置端口镜像，见表 3-5 和表 3-6。

表 3-5　在以太网端口视图下配置端口镜像（一）

操作	命令	说明
进入系统视图	system-view	-
创建端口镜像组	mirroring-group *group-id* local	必选
进入镜像目的端口的以太网端口视图	interface *interface-type interface-number*	-
定义当前端口为镜像目的端口	monitor-port	必选 镜像目的端口上不能使能 LACP 及 STP
退出当前视图	quit	-
进入镜像源端口的以太网端口视图	interface *interface-type interface-number*	-
配置镜像源端口，同时指定被镜像报文的方向	mirroring-port { inbound | outbound | both }	必选
显示镜像的参数设置	display mirroring-group { all | local }	可选 display 命令可以在任意视图下执行

表 3-6　在以太网端口视图下配置端口镜像（二）

操作	命令	说明
进入系统视图	system-view	-
创建端口镜像组	mirroring-group *group-id* local	必选
进入镜像目的端口的以太网端口视图	interface *interface-type interface-number*	-

续表

操作	命令	说明
定义当前端口为镜像目的端口	**mirroring-group** *group-id* **monitor-port**	必选 镜像目的端口上不能使能 LACP 及 STP
退出当前视图	**quit**	-
进入镜像源端口的以太网端口视图	**interface** *interface-type interface-number*	-
配置镜像源端口,同时指定被镜像报文的方向	**mirroring-group** *group-id* **mirroring-port** { **both** | **inbound** | **outbound** }	必选
显示镜像的参数设置	**display mirroring-group** { **all** | **local** }	可选 **display** 命令可以在任意视图下执行

②在系统视图下配置端口镜像,见表 3-7。

表 3-7　在系统视图下配置端口镜像

操作	命令	说明
进入系统视图	**system-view**	-
创建端口镜像组	**mirroring-group** *group-id* **local**	必选
配置镜像目的端口	**mirroring-group** *group-id* **monitor-port** *monitor-port*	必选 镜像目的端口上不能使能 LACP 及 STP
配置镜像源端口,同时指定被镜像报文的方向	**mirroring-group** *group-id* **mirroring-port** *mirroring-port-list* { **both** | **inbound** | **outbound** }	必选
显示镜像的参数设置	**display mirroring-group** { **all** | **local** }	可选 **display** 命令可以在任意视图下执行

(5) 配置举例。

如果想在 SW1 的 E4 口上监视 SW1 的 E24 口的通信情况,需要将 E4 口配置为 E24 口的镜像端口。这样,所有 E24 口上的流量都会被复制一份到 E4 口,也就可以用局域网监听软件监视到所有 E24 口上的流量了。具体配置如下:

E126 端口镜像。

配置 1:

```
[sw1]mirroring-group 1 local
[sw1]int e1/0/4
[sw1-Ethernet1/0/4]monitor-port
[sw1-Ethernet1/0/4]int e1/0/24
[sw1-Ethernet1/0/24]mirroring-port both
[sw1]display mirroring-group 1
mirroring-group 1:
    type: local
    status: active
    mirroring port:
        Ethernet1/0/24    both
    monitor port: Ethernet1/0/4
```

配置 2:

```
[H3C]sys sw1
```

[sw1]mirroring-group 1 local
[sw1]int e 1/0/4
[sw1-Ethernet1/0/4]mirroring-group 1 monitor-port
[sw1-Ethernet1/0/4]int e1/0/24
[sw1-Ethernet1/0/24]mirroring-group 1 mirroring-port both
[sw1-Ethernet1/0/24]quit
[sw1]dis mirroring-group 1
mirroring-group 1:
   type: local
   status: active
   mirroring port:
      Ethernet1/0/24   both
   monitor port: Ethernet1/0/4

配置 3：

[sw1]mirroring-group 1 local
[sw1]mirroring-group 1 monitor-port e1/0/4
[sw1]mirroring-group 1 mirroring-port e1/0/24 both
[sw1]dis mirroring-group 1
mirroring-group 1:
   type: local
   status: active
   mirroring port:
      Ethernet1/0/24   both
   monitor port: Ethernet1/0/4

E328 端口镜像：

[sw1]int e1/0/4
[sw1-Ethernet1/0/4]monitor-port
  Setting the monitor port on Ethernet1/0/4 succeeded!
[sw1-Ethernet1/0/4]quit
[sw1]int e1/0/24
[sw1-Ethernet1/0/24]mirroring-port both
[sw1-Ethernet1/0/24]quit
[sw1]dis mirror
  Monitor-port:
    Ethernet1/0/4
  Mirroring-port:
    Ethernet1/0/24   both

【参考配置】

[sw1]dis cur
#
 sysname sw1
#
 mirroring-group 1 local

```
#
radius scheme system
#
domain system
#
vlan 1
#
interface Aux1/0/0
#
interface Ethernet1/0/1
#
interface Ethernet1/0/2
#
interface Ethernet1/0/3
#
interface Ethernet1/0/4
 mirroring-group 1 monitor-port
#
interface Ethernet1/0/5
#
interface Ethernet1/0/6
#
interface Ethernet1/0/7
#
interface Ethernet1/0/8
#
interface Ethernet1/0/9
#
interface Ethernet1/0/10
#
interface Ethernet1/0/11
#
interface Ethernet1/0/12
#
interface Ethernet1/0/13
#
interface Ethernet1/0/14
#
interface Ethernet1/0/15
```

#
interface Ethernet1/0/16
#
interface Ethernet1/0/17
#
interface Ethernet1/0/17
#
interface Ethernet1/0/18
#
interface Ethernet1/0/19
#
interface Ethernet1/0/20
#
interface Ethernet1/0/21
#
interface Ethernet1/0/22
#
interface Ethernet1/0/23
#
interface Ethernet1/0/24
　　mirroring-group 1 mirroring-port both
#
interface NULL0
#
user-interface aux 0
user-interface vty 0 4
#
Return

3.4　项目总结与提高

（1）写出主要项目实施规划、步骤与实训所得的主要结论。
（2）请课余时间详细研究什么环境用到端口镜像功能，如何分析端口镜像数据。

项目四 端口汇聚

4.1 项目提出

小王在某公司上班,由于公司网络业务量增加,想提高两台交换机之间的传输带宽,他应该怎么办?

4.2 项目分析

1. 项目实训目的

掌握聚合的配置命令。
理解聚合端口的报文转发方式。

2. 项目实现功能

利用端口聚合提高带宽,达到链路备份目的。

3. 项目主要应用的技术介绍

端口汇聚,将 2 个或多个物理端口组合在一起成为一条逻辑的路径从而增加在交换机和网络节点之间的带宽,将属于这几个端口的带宽合并,给端口提供一个几倍于独立端口的独享的高带宽。它是一种封装技术,它是一条点到点的链路,链路的两端可以都是交换机,也可以是交换机和路由器,还可以是主机和交换机或路由器。基于端口汇聚功能,允许交换机与交换机、交换机与路由器、主机与交换机或路由器之间通过两个或多个端口并行连接同时传输以提供更高带宽、更大吞吐量,大幅度提高整个网络的能力。

一般情况下,在没有使用端口汇聚时,百兆以太网的双绞线的这种传输介质特性决定在两个互连的普通 10M/100M 交换机的带宽仅为 100M,这样就形成了网络主干和服务器瓶颈。要达到更高的数据传输率,则需要更换传输媒介,使用千兆光纤或升级成为千兆以太网,这样虽能在带宽上能够达到千兆,但成本却非常昂贵(可能连交换机也需要一块换掉),根本不适合低成本的中小企业和学校使用。如果使用端口汇聚技术,把多个端口通过捆绑在一起来达到更高带宽,这样较好地解决了成本和性能的矛盾。

端口汇聚是在交换机和网络设备之间比较经济的增加带宽的方法,如服务器、路由器、工作站或其他交换机。这种增加带宽的方法在当单一交换机和节点之间连接不能满足负荷时是比较有效的。

端口汇聚的主要功能就是将多个物理端口(一般为2~8个)绑定为一个逻辑的通道,使其工作起来就像一个通道一样。将多个物理链路捆绑在一起后,不但提升了整个网络的带宽,而且数据还可以同时经由被绑定的多个物理链路传输,具有链路冗余的作用,在网络出现故障或其他原因断开其中一条或多条链路时,剩下的链路还可以工作。但在 VLAN 数据传输中,各个厂家使用不同的技术,例如:思科的产品是使用其 VLAN TRUNK 技术,其他厂商的产品大多支持 802.1q 协议打上 TAG 头，这样就生成了小巨人帧，需要相同端口协议来识别，小巨人帧由于大小超过了标准以太帧的 1518 字节限制，普通网卡无法识别，需要有交换机脱 TAG。

端口汇聚功能比较适合于以下方面的具体应用：

（1）端口汇聚功能用于与服务器相连，给服务器提供独享的高带宽。

（2）端口汇聚功能用于交换机之间的级联，通过牺牲端口数来给交换机之间的数据交换提供捆绑的高带宽，提高网络速度，突破网络瓶颈，进而大幅提高网络性能。

4.3 项目实施

1．项目拓扑图

端口汇聚如图 4-1 所示。

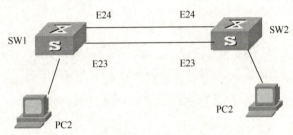

图 4-1 端口汇聚

2．项目实训环境准备

E126（2 台）、计算机（2 台）。

为了不受原来的配置影响，在实训之前先将所有的配置数据擦除后重新启动，命令为：

<h3c>Reset saved-configuration
<h3c>reboot

3．项目主要实训步骤

按照拓扑图连接所有设备。检查设备的软件版本及配置信息，所有配置为初始状态。如果配置不符合要求，请在用户视图下擦除设备中的配置文件（reset saved-configuration），然后重启设备（reboot）以使系统采用缺省的配置参数进行初始化。

（1）初始配置。

为了测试两台交换机的连通性，首先给每台交换机配置一个 3 层接口，并分配 IP 地址 192.168.1.1 /24 和 192.168.1.2 /24。然后通过 ping 命令测试互通性。

SW1 的配置：

[H3C]sys sw1
[sw1]int v
[sw1]int Vlan-interface 1
[sw1-Vlan-interface1]ip add 192.168.1.1 24

SW2 的配置：

[H3C]sys sw2
[sw2]int v 1

[sw2-Vlan-interface1]ip add 192.168.1.2 24

配置完成后，在 SW1 上 ping SW2，检查互通性：

```
[sw1]ping 192.168.1.2
  PING 192.168.1.2: 56    data bytes, press CTRL_C to break
    Reply from 192.168.1.2: bytes=56 Sequence=1 ttl=255 time=671 ms
    Reply from 192.168.1.2: bytes=56 Sequence=2 ttl=255 time=302 ms
    Reply from 192.168.1.2: bytes=56 Sequence=3 ttl=255 time=319 ms
    Reply from 192.168.1.2: bytes=56 Sequence=4 ttl=255 time=305 ms
    Reply from 192.168.1.2: bytes=56 Sequence=5 ttl=255 time=314 ms
  --- 192.168.1.2 ping statistics ---
    5 packet(s) transmitted
    5 packet(s) received
    0.00% packet loss
    round-trip min/avg/max = 302/382/671 ms
```

显然这个拓扑图会导致交换机之间存在环路，主要表现就是出现广播风暴，交换机数据转发灯不停地闪烁。广播风暴可能会导致连接在交换机上的 PC 反应速度变慢，所以在配置好端口汇聚之前，先拔掉一根交换机互联网线以消除环路。配置好汇聚后再插上。

（2）端口汇聚。

端口汇聚的前提是参加汇聚的端口必须工作在全双工方式，相同速率下，不能工作于自协商模式。

对交换机 SW1 进行配置：

```
[H3C]sysname SW1                                  //命名交换机 SW1
[SW1]int e 1/0/23                                 //进入端口 1/0/23
[SW1-Ethernet 1/0/23]speed 100                    //设置以太网端口的速率
[SW1-Ethernet 1/0/23]duplex full                  //设置以太网端口全双工状态
[SW1-Ethernet 1/0/23]port link-type trunk         //设置端口类型 trunk
[SW1-Ethernet 1/0/23]quit
[SW1] int e 1/0/24
[SW1-Ethernet 1/0/24]speed 100
[SW1-Ethernet 1/0/24]duplex full
[SW1-Ethernet 1/0/24]quit
[SW1] link-aggregation group 1 mode manual        //创建组号为 1 的手工聚合组
[SW1]int e 1/0/23
[SW1-Ethernet0/23] port link-aggregation group 1  //把端口加到组号为 1 的手工聚合组
[SW1-Ethernet0/23]quit
[SW1]int e 1/0/24
[SW1-Ethernet0/24] port link-aggregation group 1
[SW1-Ethernet0/24]quit
```

对交换机 SW2 进行配置：

```
[H3C]sysname SW2
[SW2]int e 1/0/23
[SW2-Ethernet 1/0/23]speed 100
[SW2-Ethernet 1/0/23]duplex full
```

```
[SW2-Ethernet 1/0/23]quit
[SW2] int e 1/0/24
[SW2-Ethernet 1/0/24]speed 100
[SW2-Ethernet 1/0/24]duplex full
[SW2-Ethernet 1/0/24]quit
[SW2] link-aggregation group 1 mode manual
[SW2]int e 1/0/23
[SW2-Ethernet0/23] port link-aggregation group 1
[SW2-Ethernet0/23]quit
[SW2]int e 1/0/24
[SW2-Ethernet0/24] port link-aggregation group 1
[SW2-Ethernet0/24]quit
```

现在可以把另外一根网线连接上,作为验证,可以先从 PC1 ping PC2,检查连通性,并观察是否有广播风暴。如果仍然存在请仔细检查相关配置。

```
C:\Documents and Settings\Owner>ping 192.168.1.129
Pinging 192.168.1.129 with 32 bytes of data:
Reply from 192.168.1.129: bytes=32 time<1ms TTL=128
Reply from 192.168.1.129: bytes=32 time<1ms TTL=128
Reply from 192.168.1.129: bytes=32 time<1ms TTL=128
Reply from 192.168.1.129: bytes=32 time<1ms TTL=128
Ping statistics for 192.168.1.129:
    Packets: Sent = 4, Received = 4, Lost = 0 (0% loss),
Approximate round trip times in milli-seconds:
    Minimum = 0ms, Maximum = 0ms, Average = 0ms
```

为了检查配置是否正确,可以使用 display link-aggregation 命令来查看端口汇聚情况:

```
[sw1]dis link-aggregation summary
Aggregation Group Type:D -- Dynamic, S -- Static , M -- Manual
Loadsharing Type: Shar -- Loadsharing, NonS -- Non-Loadsharing
Actor ID: 0x8000, 000f-e254-8708
  AL  AL    Partner ID          Select Unselect Share Master
  ID  Type                      Ports  Ports    Type  Port
  -------------------------------------------------------------
  1   M     none                2      0        Shar  Ethernet1/0/23

[sw1]display link-aggregation verbose
Loadsharing Type: Shar -- Loadsharing, NonS -- Non-Loadsharing
  Flags:  A -- LACP_Activity, B -- LACP_timeout, C -- Aggregation, D -- Synchronization, E --
Collecting, F -- Distributing,G -- Defaulted, H -- Expired
  Aggregation ID: 1,   AggregationType: Manual, Loadsharing Type: Shar
  Aggregation Description:
  System ID: 0x8000, 000f-e254-8708
  Port Status: S -- Selected,   U -- Unselected
  Local:
  Port                    Status  Priority  Key     Flag
  -------------------------------------------------------------
```

Ethernet1/0/23	S	32768	1	{}
Ethernet1/0/24	S	32768	1	{}

Remote:

Actor	Partner Priority	Key	SystemID	Flag
Ethernet1/0/23	0	0	0	0x0000,0000-0000-0000 {}
Ethernet1/0/24	0	0	0	0x0000,0000-0000-0000 {}

（3）验证端口汇聚的互为备份特性。

首先查看 SW1 和 SW2 的 MAC 地址表：

[sw1]display mac-ad

MAC ADDR	VLAN ID	STATE	PORT INDEX	AGING TIME(s)
000f-e25c-40f6	1	Learned	**Ethernet1/0/23**	AGING（sw2）
001e-9001-89b1	1	Learned	**Ethernet1/0/23**	AGING（pc2）
001e-9004-0216	1	Learned	Ethernet1/0/4	AGING（pc1）

[sw2]dis mac-ad

MAC ADDR	VLAN ID	STATE	PORT INDEX	AGING TIME(s)
000f-e254-8709	1	Learned	**Ethernet1/0/23**	AGING(sw1)
001e-9004-0216	1	Learned	**Ethernet1/0/23**	AGING（pc1）
001e-9001-89b1	1	Learned	Ethernet1/0/4	AGING（pc2）

如上所示，目前 SW1 和 SW2 的数据转发是通过 23 口进行转发的，上面我们看到的端口聚合状态显示 23 口是 Master Port，这两个结果是一致的。

为了检验汇聚组的 24 口是否能备份 23 口，先手工关闭 SW1 的 E1/0/23 口，再查看交换机之间的连通性和 MAC 地址表。

[sw1]int e1/0/23
[sw1-Ethernet1/0/23]shutdown
[sw1-Ethernet1/0/23]
#Apr 2 06:24:14:354 2000 sw1 L2INF/2/PORT LINK STATUS CHANGE:- 1 - Trap 1.3.6.1.6.3.1.1.5.3: portIndex is 4227802, ifAdminStatus is 2, ifOperStatus is 2
%Apr 2 06:24:14:530 2000 sw1 L2INF/5/PORT LINK STATUS CHANGE:- 1 - Ethernet1/0/23: is DOWN

C:\Documents and Settings\Owner>ping 192.168.1.129
Pinging 192.168.1.129 with 32 bytes of data:
Reply from 192.168.1.129: bytes=32 time<1ms TTL=128
Reply from 192.168.1.129: bytes=32 time<1ms TTL=128
Reply from 192.168.1.129: bytes=32 time<1ms TTL=128
Reply from 192.168.1.129: bytes=32 time<1ms TTL=128
Ping statistics for 192.168.1.129:
    Packets: Sent = 4, Received = 4, Lost = 0 (0% loss),
Approximate round trip times in milli-seconds:
    Minimum = 0ms, Maximum = 0ms, Average = 0ms

[sw1]display mac-ad

MAC ADDR	VLAN ID	STATE	PORT INDEX	AGING TIME(s)
000f-e25c-40f6	1	Learned	**Ethernet1/0/24**	AGING(**sw2**)
001e-9001-89b1	1	Learned	**Ethernet1/0/24**	AGING(**pc2**)
001e-9004-0216	1	Learned	Ethernet1/0/4	AGING(**pc1**)

[sw2]dis mac-ad

MAC ADDR	VLAN ID	STATE	PORT INDEX	AGING TIME(s)
000f-e254-8709	1	Learned	**Ethernet1/0/24**	AGING(**sw1**)
001e-9004-0216	1	Learned	**Ethernet1/0/24**	AGING(**pc1**)
001e-9001-89b1	1	Learned	Ethernet1/0/4	AGING(**pc2**)

可以看到，SW1 和 SW2 之间依然畅通。而 SW1 的转发端口由 23 口变成 24 口。

4.4 项目总结与提高

（1）写出主要项目实施规划、步骤与实训所得的主要结论。

（2）请课余时间详细研究各个厂家的端口汇聚功能是如何实现的。

项目五　VLAN 级联静态配置

5.1 项目提出

小王在某公司上班，发现业务一部和业务二部分别连接在不同的交换机上，财务一部和财务二部也是同样情况，他知道可以划分 VLAN 隔离部门间通信，但是多个交换机之间如何实现 VLAN 静态划分呢？

5.2 项目分析

1. 项目实训目的

掌握 VLAN 级联的基本配置。

2. 项目实现功能

实现多交换机 VLAN 划分。

3. 项目主要应用的技术介绍

(1) VLAN 标签。

VLAN 标签用来指示 VLAN 的成员，它封装在能够穿越局域网的帧里。这些标签在数据包进入 VLAN 的某一个交换机端口的时候被加上，在从 VLAN 的另一个端口出去的时候被去除。根据 VLAN 的端口类型会决定是给帧加入还是去除标签。VLAN 中的两类接口类型是指接入端口和骨干端口。

(2) 接入端口。

接入端口用在帧接入或者离开 VLAN 时。当接入端口收到一个帧的时候，帧并没有包含一个 VLAN 标签。一旦帧进入接入端口，会给帧加入 VLAN 标签。

当帧在交换机里面的时候，附着进入接入端口时被附上的 VLAN 标签。当帧通过目的接入端口离开交换机的时候，VLAN 标签就被去除了。传输设备和接收设备并不知道收到的帧曾经被加过 VLAN 标签。

(3) 骨干端口。

网络中包含多于一个交换机的时候，必须把 VLAN 标签的帧从一个交换机传到另一个交换机。骨干端口和接入端口的区别是骨干端口在传出帧之前，不会去除 VLAN 的标签。保留了 VLAN 标签，接收交换机就能知道传输帧属于哪一个 VLAN。帧就可以传送到接收交换机的合适端口。

(4) VLAN 标签技术。

每一个 VLAN 标记帧包含指明自身所属 VLAN 的字段。有两种主要的 VLAN 标签格式，思科公司的 Inter-Swith Link（ISL）格式和标准的 802.1Q 格式。

5.3 项目实施

1. 项目拓扑图

VLAN 级联静态配置（1）如图 5-1 所示。

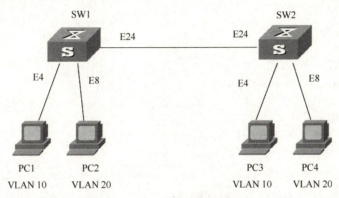

图 5-1　VLAN 级联静态配置（1）

2. 项目实训环境准备

E126（2 台）、计算机（4 台）。

为了不受原来的配置影响，在实训之前先将所有的配置数据擦除后重新启动，命令为：

<h3c>Reset saved-configuration
<h3c>reboot

3. 项目主要实训步骤

按照拓扑图连接所有设备。检查设备的软件版本及配置信息，所有配置为初始状态。如果配置不符合要求，请在用户视图下擦除设备中的配置文件（reset saved-configuration），然后重启设备（reboot）以使系统采用缺省的配置参数进行初始化。

本实训主要目的是掌握 VLAN 的基本配置。在完成 VLAN 的相关配置后，要求能够达到同一 VLAN 内的 PC 可以互通，不同 VLAN 间的 PC 不能互通的目的。

（1）初始配置。

配置表见表 5-1。

表 5-1　配置表

PC1	10.1.1.1/24	
PC2	10.1.1.2/24	
PC3	10.1.1.3/24	
PC4	10.1.1.4/24	
VLAN10	SW1:E1～E6	SW2:E1～E6
VLAN20	SW1:E7～E12	SW2:E7～E12

实际实训中，可以用两台 PC 模拟四台 PC。

（2）配置交换机 VLAN。

SW1 的配置：

[H3C]sys sw1
[sw1]vlan 10
[sw1-vlan10]port e1/0/1 to e1/0/6
[sw1-vlan10]vlan 20
[sw1-vlan20]port e1/0/7 to e1/0/12
[sw1-vlan20]quit
[sw1]

SW2 的配置：

[H3C]sys sw2
[sw2]vlan 10
[sw2-vlan10]port e1/0/1 to e1/0/6
[sw2-vlan10]vlan 20
[sw2-vlan20]port e1/0/7 to e1/0/12
[sw2-vlan20]quit
[sw2]

（3）配置交换机之间的端口为 TRUNK 口，并且允许所有 VLAN 通过

SW1 的配置：

[sw1]int e1/0/24
[sw1-Ethernet1/0/24]port link-type trunk
[sw1-Ethernet1/0/24]port trunk permit vlan all
 Please wait... Done.

SW2 的配置：

[sw2]int e1/0/24
[sw2-Ethernet1/0/24]port link-type trunk
[sw2-Ethernet1/0/24]port trunk permit vlan all
 Please wait... Done.

配置完成后，同一 VLAN 内的 PC 可以互访，不同 VLAN 间的 PC 不能互访。

（4）多交换机 VLAN。

VLAN 级联静态配置（2）如图 5-2 所示。

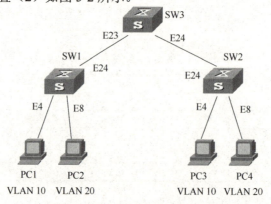

图 5-2　VLAN 级联静态配置（2）

首先，配置 SW3 的 trunk 端口：

```
[sw3]int e1/0/23
[sw3-Ethernet1/0/23]port link-type trunk
[sw3-Ethernet1/0/23]port trunk permit vlan all
[sw3]int e1/0/24
[sw3-Ethernet1/0/24]port link-type trunk
[sw3-Ethernet1/0/24]port trunk permit vlan all
```

然后在 SW3 上创建 VLAN10，VLAN20

```
[sw3]vlan 10
[sw3]vlan 20
```

思考题 1：为什么要在 SW3 上创建 VLAN10、VLAN20，而不再往里面添加任何端口。

思考题 2：如果连接了多台交换机，是否需要在每个交换机上配置 VLAN？

思考题 3：如果连接了多台交换机，每台交换机都有 10 个 VLAN，是否需要在每个交换机上配置 10 个 VLAN？

5.4 项目总结与提高

（1）写出主要项目实施规划、步骤与实训所得的主要结论。

（2）请课余时间详细研究 VLAN 加标签和去标签的具体规则。

项目六　VLAN 级联动态配置

6.1　项目提出

小王在某公司上班，发现业务一部和业务二部分别连接在不同的交换机上，财务一部和财务二部也是同样情况，他知道可以划分 VLAN 隔离部门间通信，但是多个交换机之间如何实现 VLAN 动态划分呢？

6.2　项目分析

1．项目实训目的

学习 GVRP 协议动态创建和注册 VLAN 信息。
了解 GVRP 动态注册 VLAN 的过程和端口上注册 VLAN 的三种方法。

2．项目实现功能

利用实现动态 VLAN 级联的配置。

3．项目主要应用的技术介绍

（1）GVRP。

GVRP（GARP VLAN Registration Protocol，GARP VLAN 注册协议）是 GARP 的一种应用，它基于 GARP 的工作机制，维护交换机中的 VLAN 动态注册信息，并传播该信息到其他的交换机中。

所有支持 GVRP 特性的交换机能够接收来自其他交换机的 VLAN 注册信息，并动态更新本地的 VLAN 注册信息，包括当前的 VLAN 成员、这些 VLAN 成员可以通过哪个端口到达等。而且所有支持 GVRP 特性的交换机能够将本地的 VLAN 注册信息向其他交换机传播，以便使同一交换网内所有支持 GVRP 特性的设备的 VLAN 信息达成一致。GVRP 传播的 VLAN 注册信息既包括本地手工配置的静态注册信息，也包括来自其他交换机的动态注册信息。

GVRP 在 IEEE 802.1Q 标准文本中有详细的表述。

（2）配置原则。

在配置 GVRP 时应该遵循如下原则：
- 只能在 802.1Q 兼容的端口上配置每端口 GVRP 陈述。
- 必须在 dot1Q 链路的两端都启用 GVRP。
- VLAN1 的 GVRP 注册模式始终是 fixed，并且是不可配置的，在 GVRP 启用的情况下，VLAN1 始终被 dot1Q trunk 所携带。
- VTP 修剪的情况下，它在所有禁用 GVRP 的 dot1Q TRUNK 链路上运行。

（3）注意。

不管全局下是否启用了 GVRP，都可以改变 per-trunk GVRP 配置。但是，在全局下启用 GVRP 以前，不会有任何操作。

有两种每端口 GVRP 声明方:
①在命令行下配置并存储在 NVRAM 的静态 GVRP 声明。
②在端口的实际 GVRP 的声明。

不管全局 GVRP 是否启用,只要端口在 dot1Q trunk 状态,都可以在所有的 dot1Q 兼容的端口上配置静态 GVRP 端口声明。要想使端口真正成为一个 GVRP 操作者。则必须要在全局下启用 GVRP,并且端口必须是 dot1Q trunk 口(通过命令行配置或通过动态 TRUNK 协议 DTP 协商而成)。

6.3 项目实施

1. 项目拓扑图

VLAN 级联动态配置如图 6-1 所示。

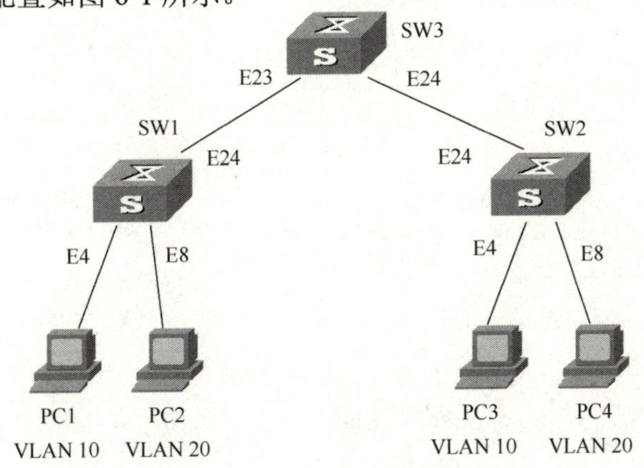

图 6-1 VLAN 级联动态配置

2. 项目实训环境准备

E126(3 台)、计算机(4 台)。

为了不受原来的配置影响,在实训之前先将所有的配置数据擦除后重新启动,命令为:

    <h3c>Reset saved-configuration
    <h3c>reboot

3. 项目主要实训步骤

本实训主要目的是掌握 VLAN 的动态配置。通过上面的 VLAN 级联静态配置,可以看到如果多台交换机相连,VLAN 的数量又很多,需要配置很多次 VLAN,工作量会很大。本次实训就是为了解决这个问题。

按照拓扑图连接所有设备。检查设备的软件版本及配置信息,所有配置为初始状态。如果配置不符合要求,请在用户视图下擦除设备中的配置文件(reset saved-configuration),然后重启设备(reboot)以使系统采用缺省的配置参数进行初始化。

(1)初始配置。

实际实训中,可以用两台 PC 模拟四台 PC。

配置表见表 6-1。

表 6-1　配置表

PC1	10.1.1.1/24	
PC2	10.1.1.2/24	
PC3	10.1.1.3/24	
PC4	10.1.1.4/24	
VLAN10	SW1:E1~E6	SW2:E1~E6
VLAN20	SW1:E7~E12	SW2:E7~E12

（2）配置交换机之间的端口为 TRUNK 口，并且允许所有 VLAN 通过。

SW1 的配置：

  [sw1]int e1/0/24
  [sw1-Ethernet1/0/24]port link-type trunk
  [sw1-Ethernet1/0/24]port trunk permit vlan all
   Please wait... Done.

SW2 的配置：

  [sw2]int e1/0/24
  [sw2-Ethernet1/0/24]port link-type trunk
  [sw2-Ethernet1/0/24]port trunk permit vlan all
   Please wait... Done.

  [sw3]int e1/0/23
  [sw3-Ethernet1/0/23]port link-type trunk
  [sw3-Ethernet1/0/23]port trunk permit vlan all
  [sw3]int e1/0/24
  [sw3-Ethernet1/0/24]port link-type trunk
  [sw3-Ethernet1/0/24]port trunk permit vlan all

（3）配置交换机 VLAN。

SW1 的配置：

  [H3C]sys sw1
  [sw1]vlan 10
  [sw1-vlan10]port e1/0/1 to e1/0/6
  [sw1-vlan10]vlan 20
  [sw1-vlan20]port e1/0/7 to e1/0/12
  [sw1-vlan20]quit
  [sw1]

SW2 的配置：

  [H3C]sys sw2
  [sw2]vlan 10
  [sw2-vlan10]port e1/0/1 to e1/0/6
  [sw2-vlan10]vlan 20
  [sw2-vlan20]port e1/0/7 to e1/0/12
  [sw2-vlan20]quit

项目六 VLAN 级联动态配置

[sw2]

（4）GVRP 配置。
SW3 的配置：
在系统视图下启动 GVRP 协议。

  [sw3]gvrp
   GVRP is enabled globally.

在每一个 TRUCK 端口启动 GVRP 协议，所以必须先配置 TRUCK 端口。

  [sw3-Ethernet1/0/23]gvrp
   GVRP is enabled on port Ethernet1/0/23.
  [sw3-Ethernet1/0/24]gvrp
   GVRP is enabled on port Ethernet1/0/24.

SW1 和 SW2 需要进行同样的 GVRP 配置，即在系统视图下和接口视图下分别启动 GVRP 协议。

  [sw1]gvrp
   GVRP is enabled globally.
  [sw1]int e1/0/24
  [sw1-Ethernet1/0/24]gvrp
   GVRP is enabled on port Ethernet1/0/24.

  [sw2]gvrp
   GVRP is enabled globally.
  [sw2]int e1/0/24
  [sw2-Ethernet1/0/24]gvrp
   GVRP is enabled on port Ethernet1/0/24.

配置完成后，同一 VLAN 内的 PC 可以互访，不同 VLAN 间的 PC 不能互访。
（5）查看 GVRP 信息和 VLAN 信息。
SW3 上的信息：

```
[sw3]display gvrp statistics
        GVRP statistics on port Ethernet1/0/23
            GVRP Status                  : Enabled
            GVRP Failed Registrations    : 0
            GVRP Last Pdu Origin         : 000f-e207-f2e0
            GVRP Registration Type       : Normal
        GVRP statistics on port Ethernet1/0/24
            GVRP Status                  : Enabled
            GVRP Failed Registrations    : 0
            GVRP Last Pdu Origin         : 000f-e207-f2e0
            GVRP Registration Type       : Normal
[sw3]display vlan
  The following VLANs exist:
    1(default), 10, 20
```

(6) 在 SW1 上手工添加一个 VLAN30，查看 3 个交换机 VLAN 情况。
① SW1 添加 VLAN30。

    [sw1]vlan 30

② 查看配置。

SW1 的 VLAN：

    [sw1]display vlan
     The following VLANs exist:
      1(default), 10, 20, **30**
    [sw1]display vlan all
    VLAN ID: 1
    VLAN Type: static
    Route Interface: not configured
    Description: VLAN 0001
    Name: VLAN 0001
    Tagged   Ports: none
    Untagged Ports:
     Ethernet1/0/13   Ethernet1/0/14   Ethernet1/0/15
     Ethernet1/0/16   Ethernet1/0/17   Ethernet1/0/18
     Ethernet1/0/19   Ethernet1/0/20   Ethernet1/0/21
     Ethernet1/0/22   Ethernet1/0/23   Ethernet1/0/24
     GigabitEthernet1/1/1   GigabitEthernet1/1/2   GigabitEthernet1/1/3
     GigabitEthernet1/1/4
    VLAN ID: 10
    VLAN Type: static
    Route Interface: not configured
    Description: VLAN 0010
    Name: VLAN 0010
    Tagged   Ports:
     Ethernet1/0/24
    Untagged Ports:
     Ethernet1/0/1   Ethernet1/0/2   Ethernet1/0/3
     Ethernet1/0/4   Ethernet1/0/5   Ethernet1/0/6
    VLAN ID: 20
    VLAN Type: static
    Route Interface: not configured
    Description: VLAN 0020
    Name: VLAN 0020
    Tagged   Ports:
     Ethernet1/0/24
    Untagged Ports:
     Ethernet1/0/7   Ethernet1/0/8   Ethernet1/0/9
     Ethernet1/0/10   Ethernet1/0/11   Ethernet1/0/12

    VLAN ID: 30
    VLAN Type: static

Route Interface: not configured
Description: VLAN 0030
Name: VLAN 0030
Tagged   Ports:
  Ethernet1/0/24
Untagged Ports: none

SW2 的 VLAN：

[sw2]display vlan
 The following VLANs exist:
  1(default), 10, 20, 30
[sw2]display vlan all
 VLAN ID: 1
 VLAN Type: static
 Route Interface: not configured
 Description: VLAN 0001
 Name: VLAN 0001
 Tagged   Ports: none
 Untagged Ports:
  Ethernet1/0/13   Ethernet1/0/14   Ethernet1/0/15
  Ethernet1/0/16   Ethernet1/0/17   Ethernet1/0/18
  Ethernet1/0/19   Ethernet1/0/20   Ethernet1/0/21
  Ethernet1/0/22   Ethernet1/0/23   Ethernet1/0/24
 VLAN ID: 10
 VLAN Type: static
 Route Interface: not configured
 Description: VLAN 0010
 Name: VLAN 0010
 Tagged   Ports:
  Ethernet1/0/24
 Untagged Ports:
  Ethernet1/0/1   Ethernet1/0/2   Ethernet1/0/3
  Ethernet1/0/4   Ethernet1/0/5   Ethernet1/0/6
 VLAN ID: 20
 VLAN Type: static
 Route Interface: not configured
 Description: VLAN 0020
 Name: VLAN 0020
 Tagged   Ports:
  Ethernet1/0/24
 Untagged Ports:
  Ethernet1/0/7   Ethernet1/0/8   Ethernet1/0/9
  Ethernet1/0/10   Ethernet1/0/11   Ethernet1/0/12
 VLAN ID: 30
 VLAN Type: dynamic
 Route Interface: not configured
 Description: VLAN 0030

Name: VLAN 0030
　Tagged　Ports:
　　Ethernet1/0/24
　Untagged Ports: none

SW3 的 VLAN：

[sw3]display vlan
　The following VLANs exist:
　　1(default), 10, 20, 30
[sw3]display vlan all
VLAN ID: 1
VLAN Type: static
Route Interface: not configured
Description: VLAN 0001
Name: VLAN 0001
　Tagged　Ports: none
Untagged Ports:
　Ethernet1/0/1　　　　　Ethernet1/0/2　　　　　Ethernet1/0/3
　Ethernet1/0/4　　　　　Ethernet1/0/5　　　　　Ethernet1/0/6
　Ethernet1/0/7　　　　　Ethernet1/0/8　　　　　Ethernet1/0/9
　Ethernet1/0/10　　　　Ethernet1/0/11　　　　Ethernet1/0/12
　Ethernet1/0/13　　　　Ethernet1/0/14
　Ethernet1/0/16　　　　Ethernet1/0/17　　　　Ethernet1/0/18
　Ethernet1/0/19　　　　Ethernet1/0/20　　　　Ethernet1/0/21
　Ethernet1/0/22　　　　Ethernet1/0/23　　　　Ethernet1/0/24
　GigabitEthernet1/1/1　GigabitEthernet1/1/2　GigabitEthernet1/1/3
　GigabitEthernet1/1/4

VLAN ID: 10
VLAN Type: dynamic
Route Interface: not configured
Description: VLAN 0010
Name: VLAN 0010
　Tagged　Ports:
　　Ethernet1/0/23　　　　Ethernet1/0/24
　Untagged Ports: none

VLAN ID: 20
VLAN Type: dynamic
Route Interface: not configured
Description: VLAN 0020
Name: VLAN 0020
　Tagged　Ports:
　　Ethernet1/0/23　　　　Ethernet1/0/24
　Untagged Ports: none

VLAN ID: 30
VLAN Type: dynamic
Route Interface: not configured
Description: VLAN 0030

Name: VLAN 0030
Tagged   Ports:
  Ethernet1/0/24
Untagged Ports: none

（7）查看 TRUNK 接口。

SW1：

[sw1]display interface Ethernet 1/0/24
Ethernet1/0/24 current state : UP
IP Sending Frames' Format is PKTFMT_ETHNT_2, Hardware address is 000f-e254-8708
Media type is twisted pair, loopback not set
Port hardware type is 100_BASE_TX
100Mbps-speed mode, full-duplex mode
Link speed type is autonegotiation, link duplex type is autonegotiation
Flow-control is not enabled
The Maximum Frame Length is 9216
Broadcast MAX-ratio: 100%
Unicast MAX-ratio: 100%
Multicast MAX-ratio: 100%
Allow jumbo frame to pass
PVID: 1
Mdi type: auto
Port link-type: trunk
  VLAN passing   : 1(default vlan), 10, 20, 30
  VLAN permitted: 1(default vlan), 2-4094
  Trunk port encapsulation: IEEE 802.1q
Last 300 seconds input: 1 packets/sec 112 bytes/sec
Last 300 seconds output: 0 packets/sec 40 bytes/sec
Input(total):   268004 packets, 26162021 bytes
          265756 broadcasts, 408 multicasts, 0 pauses
Input(normal):   - packets, - bytes
          - broadcasts, - multicasts, - pauses
Input: 1 input errors, 0 runts, 0 giants,   - throttles, 0 CRC
          0 frame,   - overruns, 1 aborts, 0 ignored, - parity errors
Output(total): 7723513 packets, 997096040 bytes
          7722727 broadcasts, 632 multicasts, 0 pauses
Output(normal): - packets, - bytes
          - broadcasts, - multicasts, - pauses
Output: 0 output errors,   - underruns, - buffer failures
          0 aborts, 0 deferred, 0 collisions, 0 late collisions
          0 lost carrier, - no carrier

SW2：

[sw2]display interface Ethernet 1/0/24
Ethernet1/0/24 current state : UP
IP Sending Frames' Format is PKTFMT_ETHNT_2, Hardware address is 000f-e25c-40f6
Media type is twisted pair, loopback not set

Port hardware type is 100_BASE_TX
100Mbps-speed mode, full-duplex mode
Link speed type is autonegotiation, link duplex type is autonegotiation
Flow-control is not enabled
The Maximum Frame Length is 1536
Broadcast MAX-ratio: 100%
PVID: 1
Mdi type: normal
Port link-type: trunk
 VLAN passing   : 1(default vlan), 10, 20, 30
 VLAN permitted: 1(default vlan), 2-4094
 Trunk port encapsulation: IEEE 802.1q
Last 300 seconds input: 0 packets/sec 25 bytes/sec
Last 300 seconds output: 1 packets/sec 107 bytes/sec
Input(total):   7317661 packets, 900379013 bytes
        7316759 broadcasts, 748 multicasts, 0 pauses
Input(normal):   7317661 packets, 900379013 bytes
        7316759 broadcasts, 748 multicasts, 0 pauses
Input: 0 input errors, 0 runts, 0 giants,   - throttles, 0 CRC
         0 frame,   - overruns, 0 aborts, - ignored, - parity errors
Output(total): 268484 packets, 26212671 bytes
       266321 broadcasts, 324 multicasts, 0 pauses
Output(normal): 268484 packets, - bytes
       266321 broadcasts, 324 multicasts, - pauses
Output: 0 output errors,   - underruns, - buffer failures
        0 aborts, 0 deferred, 0 collisions, 0 late collisions
        - lost carrier, - no carrier

SW3：

[sw3]display interface Ethernet 1/0/23
Ethernet1/0/23 current state : UP
IP Sending Frames' Format is PKTFMT_ETHNT_2, Hardware address is 000f-e254-8702

Media type is twisted pair, loopback not set
Port hardware type is 100_BASE_TX
100Mbps-speed mode, full-duplex mode
Link speed type is autonegotiation, link duplex type is autonegotiation
Flow-control is not enabled
The Maximum Frame Length is 9216
Broadcast MAX-ratio: 100%
Unicast MAX-ratio: 100%
Multicast MAX-ratio: 100%
Allow jumbo frame to pass
PVID: 1
Mdi type: auto
Port link-type: trunk
 VLAN passing   : 1(default vlan), 10, 20, 30

VLAN permitted: 1(default vlan), 2-4094
Trunk port encapsulation: IEEE 802.1q
Last 300 seconds input: 1 packets/sec 107 bytes/sec
Last 300 seconds output: 0 packets/sec 26 bytes/sec
Input(total):   1911 packets, 188320 bytes
          1731 broadcasts, 180 multicasts, 0 pauses
Input(normal):   - packets, - bytes
          - broadcasts, - multicasts, - pauses
Input: 0 input errors, 0 runts, 0 giants,   - throttles, 0 CRC
          0 frame,   - overruns, 0 aborts, 0 ignored, - parity errors
Output(total): 126193 packets, 29206721 bytes
          125830 broadcasts, 362 multicasts, 0 pauses
Output(normal): - packets, - bytes
          - broadcasts, - multicasts, - pauses
Output: 0 output errors,   - underruns, - buffer failures
          0 aborts, 0 deferred, 0 collisions, 0 late collisions
          0 lost carrier, - no carrier
[sw3]display interface Ethernet 1/0/24
Ethernet1/0/24 current state : UP
IP Sending Frames' Format is PKTFMT_ETHNT_2, Hardware address is 000f-e254-8702
Media type is twisted pair, loopback not set
Port hardware type is 100_BASE_TX
100Mbps-speed mode, full-duplex mode
Link speed type is autonegotiation, link duplex type is autonegotiation
Flow-control is not enabled
The Maximum Frame Length is 9216
Broadcast MAX-ratio: 100%
Unicast MAX-ratio: 100%
Multicast MAX-ratio: 100%
Allow jumbo frame to pass
PVID: 1
Mdi type: auto
Port link-type: trunk
  VLAN passing   : 1(default vlan), 10, 20
VLAN permitted: 1(default vlan), 2-4094
Trunk port encapsulation: IEEE 802.1q
Last 300 seconds input: 0 packets/sec 26 bytes/sec
Last 300 seconds output: 0 packets/sec 87 bytes/sec
Input(total):   507981 packets, 120196193 bytes
          507691 broadcasts, 288 multicasts, 0 pauses
Input(normal):   - packets, - bytes
          - broadcasts, - multicasts, - pauses
Input: 1 input errors, 0 runts, 0 giants,   - throttles, 0 CRC
          0 frame,   - overruns, 1 aborts, 0 ignored, - parity errors
Output(total): 1663 packets, 160014 bytes
          1351 broadcasts, 312 multicasts, 0 pauses
Output(normal): - packets, - bytes

```
                - broadcasts, - multicasts, - pauses
     Output: 0 output errors,    - underruns, - buffer failures
                0 aborts, 0 deferred, 0 collisions, 0 late collisions
                0 lost carrier, - no carrier
```

这时看到，SW1 的 TRUNK 端口（24 口）允许 VLAN30 通过，同样 SW2 的 E24 口和 SW3 的 E23 口也允许 VLAN30 通过，但是 SW3 的 E24 口却只允许 VLAN1、VLAN10、VLAN20 通过，VLAN30 不能通过。

实际上，这是 GVRP 进行 VLAN 注册的一条规则，由于交换机 SW2 上没有手工配置 VLAN30，也就是说 SW2 上没有 VLAN30 这个成员，所以 SW2 和 SW3 之间的 TRUNK 链路不应该让 VLAN30 的帧通过。

如果我们在 SW2 上手工创建 VLAN30，SW3 的 E24 就会允许 VLAN30 通过。下面验证一下：

```
[sw2]vlan 30
 Dynamic VLAN is configured, now changed to static!
[sw3]display interface Ethernet 1/0/24
 Ethernet1/0/24 current state : UP
 IP Sending Frames' Format is PKTFMT_ETHNT_2, Hardware address is 000f-e254-8702
 Media type is twisted pair, loopback not set
 Port hardware type is 100_BASE_TX
 100Mbps-speed mode, full-duplex mode
 Link speed type is autonegotiation, link duplex type is autonegotiation
 Flow-control is not enabled
 The Maximum Frame Length is 9216
 Broadcast MAX-ratio: 100%
 Unicast MAX-ratio: 100%
 Multicast MAX-ratio: 100%
 Allow jumbo frame to pass
 PVID: 1
 Mdi type: auto
 Port link-type: trunk
  VLAN passing    : 1(default vlan), 10, 20, 30
  VLAN permitted: 1(default vlan), 2-4094
  Trunk port encapsulation: IEEE 802.1q
 Last 300 seconds input:   1 packets/sec 101 bytes/sec
 Last 300 seconds output:    0 packets/sec 18 bytes/sec
 Input(total):   508742 packets, 120269953 bytes
                508350 broadcasts, 390 multicasts, 0 pauses
 Input(normal):   - packets, - bytes
                - broadcasts, - multicasts, - pauses
 Input:   1 input errors, 0 runts, 0 giants,   - throttles, 0 CRC
                0 frame,   - overruns, 1 aborts, 0 ignored, - parity errors
 Output(total): 1663 packets, 160014 bytes
                1351 broadcasts, 312 multicasts, 0 pauses
 Output(normal): - packets, - bytes
                - broadcasts, - multicasts, - pauses
```

```
        Output: 0 output errors,   - underruns, - buffer failures
                0 aborts, 0 deferred, 0 collisions, 0 late collisions
                0 lost carrier, - no carrier
```

（8）配置 GVRP 注册方法。

在交换机的 TRUNK 端口上，VLAN 的 GVRP 注册方法有三种：normal、fixed 和 forbidden。其中 normal 是默认的注册方法，表示允许在该端口手工或动态创建、注册和注销 VLAN。Fixed 表示允许在该端口手工创建和注册 VLAN，不允许动态注册或注销 VLAN。Forbidden 表示在端口注销除 VLAN1 之外的所有 VLAN，并禁止在该端口创建和注册任何其他 VLAN。下面举例说明其他两个选项。

Fixed 举例：

在 SW1 的 E24 口上配置 fixed 方法，在 SW2 上创建 VLAN40，查看 SW1 和 SW3 的 VLAN 信息和 TRUNK 端口状态。

```
[sw1]interface Ethernet 1/0/24
[sw1-Ethernet1/0/24]gvrp registration fixed
  Please wait............................................ Done.

[sw2]vlan 40
[sw2-vlan40]quit
[sw2]display vlan
  The following VLANs exist:
    1(default), 10, 20, 30, 40

[sw3]display vlan
  The following VLANs exist:
    1(default), 10, 20, 30, 40

[sw1]display vlan
  The following VLANs exist:
    1(default), 10, 20, 30

[sw1]display interface Ethernet 1/0/24
  Ethernet1/0/24 current state : UP
  IP Sending Frames' Format is PKTFMT_ETHNT_2, Hardware address is 000f-e254-8708

  Media type is twisted pair, loopback not set
  Port hardware type is 100_BASE_TX
  100Mbps-speed mode, full-duplex mode
  Link speed type is autonegotiation, link duplex type is autonegotiation
  Flow-control is not enabled
  The Maximum Frame Length is 9216
  Broadcast MAX-ratio: 100%
  Unicast MAX-ratio: 100%
  Multicast MAX-ratio: 100%
  Allow jumbo frame to pass
  PVID: 1
```

　　　　　Mdi type: auto
　　　　　Port link-type: trunk
　　　　 VLAN passing　: 1(default vlan), 10, 20, 30
　　　　　VLAN permitted: 1(default vlan), 2-4094
　　　　　Trunk port encapsulation: IEEE 802.1q
　　　　Last 300 seconds input:　1 packets/sec 120 bytes/sec
　　　　Last 300 seconds output:　0 packets/sec 26 bytes/sec
　　　　Input(total):　270365 packets, 26389218 bytes
　　　　　　　267706 broadcasts, 806 multicasts, 0 pauses
　　　　Input(normal):　- packets, - bytes
　　　　　　　- broadcasts, - multicasts, - pauses
　　　　Input:　2 input errors, 0 runts, 0 giants,　- throttles, 0 CRC
　　　　　　　0 frame,　- overruns, 2 aborts, 0 ignored, - parity errors
　　　　Output(total): 7724028 packets, 997136793 bytes
　　　　　　　7722818 broadcasts, 1043 multicasts, 0 pauses
　　　　Output(normal): - packets, - bytes
　　　　　　　- broadcasts, - multicasts, - pauses
　　　　Output: 0 output errors,　- underruns, - buffer failures
　　　　　　　0 aborts, 0 deferred, 0 collisions, 0 late collisions
　　　　　　　0 lost carrier, - no carrier

Forbidden 举例：

在 SW1 的 E24 口上配置 forbidden 方法，查看 SW1 的 VLAN 信息和 TRUNK 端口状态。

　　　[sw1]interface Ethernet 1/0/24
　　　[sw1-Ethernet1/0/24]**gvrp registration forbidden**
　　　　Please wait... Done.
　　　[sw1-Ethernet1/0/24]quit
　　　[sw1]**display vlan**
　　　　The following VLANs exist:
　　　　　1(default), 10, 20, 30
　　　[sw1]**display interface Ethernet 1/0/24**
　　　　Ethernet1/0/24 current state : UP
　　　　IP Sending Frames' Format is PKTFMT_ETHNT_2, Hardware address is 000f-e254-8708
　　　　Media type is twisted pair, loopback not set
　　　　Port hardware type is 100_BASE_TX
　　　　100Mbps-speed mode, full-duplex mode
　　　　Link speed type is autonegotiation, link duplex type is autonegotiation
　　　　Flow-control is not enabled
　　　　The Maximum Frame Length is 9216
　　　　Broadcast MAX-ratio: 100%
　　　　Unicast MAX-ratio: 100%
　　　　Multicast MAX-ratio: 100%
　　　　Allow jumbo frame to pass
　　　　PVID: 1
　　　　Mdi type: auto
　　　　Port link-type: trunk
　　　　 VLAN passing　: 1(default vlan)

```
       VLAN permitted: 1(default vlan), 2-4094
       Trunk port encapsulation: IEEE 802.1q
    Last 300 seconds input:    1 packets/sec 108 bytes/sec
    Last 300 seconds output:   0 packets/sec 19 bytes/sec
    Input(total):   270583 packets, 26408712 bytes
              267845 broadcasts, 885 multicasts, 0 pauses
    Input(normal):   - packets, - bytes
              - broadcasts, - multicasts, - pauses
    Input:   2 input errors, 0 runts, 0 giants,   - throttles, 0 CRC
              0 frame,  - overruns, 2 aborts, 0 ignored, - parity errors
    Output(total): 7724073 packets, 997140075 bytes
              7722818 broadcasts, 1088 multicasts, 0 pauses
    Output(normal): - packets, - bytes
              - broadcasts, - multicasts, - pauses
    Output: 0 output errors,   - underruns, - buffer failures
              0 aborts, 0 deferred, 0 collisions, 0 late collisions
              0 lost carrier, - no carrier
```

6.4 项目总结与提高

(1) 写出主要项目实施规划、步骤与实训所得的主要结论。

(2) 请课余时间详细研究 GVRP 协议。

项目七　VLAN 的 Hybrid 端口配置

7.1　项目提出

小王在某公司上班，业务一部和业务二部分别连接在不同的交换机上，财务一部和财务二部也是同样情况，如果需要财务一部和财务二部可以通信，而业务部分需要隔离，他该怎么办？

7.2　项目分析

1．项目实训目的

掌握 Hybrid 端口的基本特性，了解其常见应用。

2．项目实现功能

利用 Hybrid 端口属性有选择性地隔离 VLAN。

3．项目主要应用的技术介绍

Hybrid 接口也能够允许多个 VLAN 帧通过并且还可以指定哪些 VLAN 数据帧被剥离标签。

Hybrid 接口是华为、H3C 交换机的一种端口模式，和 Trunk 接口一样在设置允许指定的 VLAN 通过 Hybrid 端口之前，该 VLAN 必须已经存在。

Hybrid 端口和 Trunk 端口在接收数据时，处理思路方法是一样的，唯一区别之处在于发送数据时，Hybrid 端口具有解除多 VLAN 标签的功能，Hybrid 端口可以允许多个 VLAN 的报文发送时不打标签，从而增加了网络的灵活性，在一定程度上也增加了安全性，而 Trunk 端口只允许缺省 VLAN 的报文发送时不打标签。

需要注意的是 Hybrid 端口和 Trunk 端口不能直接切换，只能先设为 Access 端口，再设置为其他类型端口。基于 MAC 地址、协议、IP 子网的 VLAN 只对 Hybrid 端口配置有效。

7.3　项目实施

1．项目拓扑图

VLAN 的 Hybrid 端口配置如图 7-1 所示。

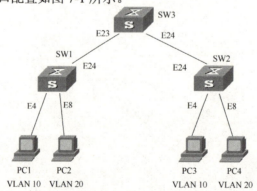

图 7-1　VLAN 的 Hybrid 端口配置

2. 项目实训环境准备

E126（3台）、计算机（4台）。

为了不受原来的配置影响，在实训之前先将所有的配置数据擦除后重新启动，命令为：

    <h3c>Reset saved-configuration
    <h3c>reboot

3. 项目主要实训步骤

本实训主要目的是配置交换机之间的端口为 Hybrid 端口，允许 VLAN10 和 VLAN20 的帧通过 Hybrid 端口，VLAN10 的帧在 Hybrid 端口打上 VLAN10 的 tag（标签），VLAN20 的帧不打 tag。

按照拓扑图连接所有设备。检查设备的软件版本及配置信息，所有配置为初始状态。如果配置不符合要求，请在用户视图下擦除设备中的配置文件（reset saved-configuration），然后重启设备（reboot）以使系统采用缺省的配置参数进行初始化。

如果交换机上有 Trunk 端口存在时不能配置 Hybrid 端口，所以需要先把 Trunk 端口改成 Access 端口，注意不能直接把 Trunk 端口改成 Hybrid 端口。

[H3C-Ethernet1/0/24]port link-type access

（1）初始配置。

配置表见表 7-1。

表 7-1 配置表

PC1	10.1.1.1/24	
PC2	10.1.1.2/24	
PC3	10.1.1.3/24	
PC4	10.1.1.4/24	
VLAN10	SW1:E1~E6	SW2:E1~E6
VLAN20	SW1:E7~E12	SW2:E7~E12

实际实训中，可以用两台 PC 模拟四台 PC。

SW1 的配置：

    [H3C]sys sw1
    [sw1]vlan 10
    [sw1-vlan10]port e1/0/1 to e1/0/6
    [sw1-vlan10]vlan 20
    [sw1-vlan20]port e1/0/7 to e1/0/12
    [sw1-vlan20]quit
    [sw1]

SW2 的配置：

    [H3C]sys sw2
    [sw2]vlan 10
    [sw2-vlan10]port e1/0/1 to e1/0/6
    [sw2-vlan10]vlan 20
    [sw2-vlan20]port e1/0/7 to e1/0/12

[sw2-vlan20]quit
[sw2]

SW3 的配置：

[sw3]vlan 10
[sw3-vlan10]vlan 20

（2）配置交换机之间的端口为 Hybrid 口。

SW1 的配置：

[sw1]int e1/0/24
[sw1-Ethernet1/0/24]port link-type hybrid

SW2 的配置：

[sw2]int e1/0/24
[sw2-Ethernet1/0/24]port link-type hybrid

[sw3]int e1/0/23
[sw3-Ethernet1/0/23]port link-type hybrid
[sw3]int e1/0/24
[sw3-Ethernet1/0/24]port link-type hybrid

（3）配置 Hybrid 端口，VLAN10 打 tag，VLAN20 不打 tag。

首先看到 SW1 的 E24 口属性为 Hybrid，PVID 为 1，tagged VLAN ID 没有配置，而 untagged VLAN ID 默认为 1。

```
[sw1]display interface Ethernet 1/0/24
  Ethernet1/0/24 current state : UP
  IP Sending Frames' Format is PKTFMT_ETHNT_2, Hardware address is 000f-e25c-40f6

  Media type is twisted pair, loopback not set
  Port hardware type is 100_BASE_TX
  100Mbps-speed mode, full-duplex mode
  Link speed type is autonegotiation, link duplex type is autonegotiation
  Flow-control is not enabled
  The Maximum Frame Length is 1536
  Broadcast MAX-ratio: 100%
  PVID: 1
  Mdi type: normal
  Port link-type: hybrid
   Tagged    VLAN ID : none
   Untagged VLAN ID : 1
  Last 300 seconds input:    0 packets/sec 4 bytes/sec
  Last 300 seconds output:   0 packets/sec 2 bytes/sec
  Input(total):   152 packets, 14824 bytes
          99 broadcasts, 53 multicasts, 0 pauses
  Input(normal):  152 packets, 14824 bytes
          99 broadcasts, 53 multicasts, 0 pauses
```

```
        Input:   0 input errors, 0 runts, 0 giants,   - throttles, 0 CRC
                 0 frame,   - overruns, 0 aborts, - ignored, - parity errors
        Output(total): 1730 packets, 168556 bytes
                 1685 broadcasts, 45 multicasts, 0 pauses
```

然后，配置 SW1、SW2、SW3，VLAN10 打 tag，VLAN20 不打 tag。

```
[sw1-Ethernet1/0/24]port hybrid vlan 10 tagged
[sw1-Ethernet1/0/24]port hybrid vlan 20 untagged

[sw2-Ethernet1/0/24]port hybrid vlan 10 tagged
[sw2-Ethernet1/0/24]port hybrid vlan 20 untagged


[sw3]int e1/0/23
[sw3-Ethernet1/0/23]port lin
[sw3-Ethernet1/0/23]port link-t hy
[sw3-Ethernet1/0/23]port hy v 10 t
[sw3-Ethernet1/0/23]port hy v 20 u
[sw3-Ethernet1/0/23]int e1/0/24
[sw3-Ethernet1/0/24]port lin
[sw3-Ethernet1/0/24]port link-t hy
[sw3-Ethernet1/0/24]port hy v 10 t
[sw3-Ethernet1/0/24]port hy v 20 u
```

(4) 查看端口信息。

```
IP Sending Frames' Format is PKTFMT_ETHNT_2, Hardware address is 000f-e254-8708
        Media type is twisted pair, loopback not set
        Port hardware type is 100_BASE_TX
        100Mbps-speed mode, full-duplex mode
        Link speed type is autonegotiation, link duplex type is autonegotiation
        Flow-control is not enabled
        The Maximum Frame Length is 9216
        Broadcast MAX-ratio: 100%
        Unicast MAX-ratio: 100%
        Multicast MAX-ratio: 100%
        Allow jumbo frame to pass
        PVID: 1
        Mdi type: auto
        Port link-type: hybrid
         Tagged     VLAN ID : 10
         Untagged VLAN ID : 1, 20
        Last 300 seconds input:    0 packets/sec 3 bytes/sec
        Last 300 seconds output:   0 packets/sec 25 bytes/sec
        Input(total):   1832 packets, 183042 bytes
                  1720 broadcasts, 107 multicasts, 0 pauses
        Input(normal):   - packets, - bytes
                 - broadcasts, - multicasts, - pauses
        Input:   0 input errors, 0 runts, 0 giants,   - throttles, 0 CRC
```

```
           Input:    0 input errors, 0 runts, 0 giants,   - throttles, 0 CRC
                     0 frame,   - overruns, 0 aborts, 0 ignored, - parity errors
           Output(total): 859 packets, 89556 bytes
                     463 broadcasts, 131 multicasts, 0 pauses
           Output(normal): - packets, - bytes
                     - broadcasts, - multicasts, - pauses
           Output: 0 output errors,   - underruns, - buffer failures
                     0 aborts, 0 deferred, 0 collisions, 0 late collisions
                     0 lost carrier, - no carrier
```

Hybrid 端口是介于 Access 端口和 Trunk 端口之间的一种端口属性，通过 Hybrid 端口的设置，可以配置部分 VLAN 的帧可以打上 VLAN 标记，另一部分 VLAN 在通过交换机间链路时，不打 VLAN 标记，从而实现对部分 VLAN 隔离的目的。当然，如果您把所有 VLAN 在 Hybrid 端口都配置为 untagged，那么交换机之间的所有 VLAN 都被隔离，能够满足一些特殊应用。

7.4 项目总结与提高

（1）写出主要项目实施规划、步骤与实训所得的主要结论。

（2）请课余时间详细研究华三和思科的网络设备 Hybrid 端口的属性。

项目八　VLAN 间静态路由配置

8.1　项目提出

小王在某公司上班，业务部和财务部分别在不同交换机的不同 VLAN 下，如果需要财务部和业务部可以通信，他该怎么办？

8.2　项目分析

1．项目实训目的

掌握静态路由在三层交换机上的配置。
理解 ipfdb 表的组成和作用。

2．项目实现功能

利用静态路由使跨交换机的 VLAN 间可以通信。

3．项目主要应用的技术介绍

用 VLAN 分段，隔离了 VLAN 间的通信，用支持 VLAN 的路由器（三层设备）可以建立 VLAN 间通信。但使用路由器来互联企业园区网中不同的 VLAN 显然不合时代的潮流。因为我们可以使用三层交换来实现。

两者的差异：

差别 1（性能）：传统的路由器基于微处理器转发报文，靠软件处理，而三层交换机通过 ASIC 硬件来进行报文转发，性能差别很大；

差别 2（接口类型）：三层交换机的接口基本都是以太网接口，没有路由器接口类型丰富；

差别 3：三层交换机，还可以工作在二层模式，对某些不需路由的报文直接交换，而路由器不具有二层的功能。

8.3　项目实施

1．项目拓扑图

VLAN 间静态路由配置如图 8-1 所示。

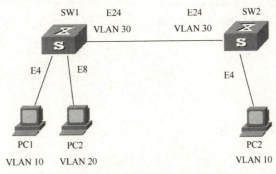

图 8-1　VLAN 间静态路由配置

2. 项目实训环境准备

E328（2 台）、计算机（3 台）。

为了不受原来的配置影响，在实训之前先将所有的配置数据擦除后重新启动，命令为：

    <h3c>Reset saved-configuration
    <h3c>reboot

3. 项目主要实训步骤

按照拓扑图连接所有设备。检查设备的软件版本及配置信息，所有配置为初始状态。如果配置不符合要求，请在用户视图下擦除设备中的配置文件（reset saved-configuration），然后重启设备（reboot）以使系统采用缺省的配置参数进行初始化。

（1）初始配置。

配置表见表 8-1。

表 8-1 配置表

PC1	VLAN10	10.1.1.2/24	网关：10.1.1.1
PC2	VLAN20	10.1.2.2/24	网关：10.1.2.1
PC3	VLAN10	10.1.3.2/24	网关：10.1.3.1
SW1:VLAN10	SW1:E1~E6	10.1.1.1/24	
SW1:VLAN20	SW1:E7~E12	10.1.2.1/24	
SW1:VLAN30		10.1.4.1/24	
SW2: VLAN10	SW2:E1~E6	10.1.3.1/24	
SW2: VLAN30		10.1.4.2/24	

（2）配置交换机 VLAN，TRUNK 端口。

SW1 的配置：

    [H3C]sys sw1
    [sw1]vlan 10
    [sw1-vlan10]port e1/0/1 t e1/0/6
    [sw1-vlan10]vlan 20
    [sw1-vlan20]port e1/0/7 t e 1/0/12
    [sw1-vlan20]vlan 30
    [sw1-vlan30]quit
    [sw1]int e1/0/24
    [sw1-Ethernet1/0/24]port link-type trunk
    [sw1-Ethernet1/0/24]port trunk permit vlan all
     Please wait... Done.

SW2 的配置：

    [H3C]sys sw2
    [sw2]vlan 10
    [sw2-vlan10]port e1/0/1 t e1/0/6
    [sw2]vlan 30
    [sw2-vlan30]int e1/0/24
    [sw2-Ethernet1/0/24]port link-type trunk

项目八 VLAN 间静态路由配置

```
[sw2-Ethernet1/0/24]port trunk permit vlan all
  Please wait............................ Done.
```

(3) 配置 PC 和交换机的网段地址。

SW1 的配置：

```
[sw1]int Vlan-interface 10
[sw1-Vlan-interface10]ip add 10.1.1.1 24
[sw1-Vlan-interface10]int v 20
[sw1-Vlan-interface20]ip add 10.1.2.1 24
[sw1-Vlan-interface20]int v 30
[sw1-Vlan-interface30]ip add 10.1.4.1 24
```

这时 10.1.1.0 网段、10.1.2.0 网段和 10.1.4.1，能够互通。请测试连通性并查看路由表。

```
[sw1]dis ip rou
  Routing Table: public net
Destination/Mask    Protocol    Pre   Cost    Nexthop      Interface
10.1.1.0/24         DIRECT      0     0       10.1.1.1     Vlan-interface10
10.1.1.1/32         DIRECT      0     0       127.0.0.1    InLoopBack0
10.1.2.0/24         DIRECT      0     0       10.1.2.1     Vlan-interface20
10.1.2.1/32         DIRECT      0     0       127.0.0.1    InLoopBack0
10.1.4.0/24         DIRECT      0     0       10.1.4.1     Vlan-interface30
10.1.4.1/32         DIRECT      0     0       127.0.0.1    InLoopBack0
127.0.0.0/8         DIRECT      0     0       127.0.0.1    InLoopBack0
127.0.0.1/32        DIRECT      0     0       127.0.0.1    InLoopBack0
```

SW2 的配置：

```
[sw2]int v 10
[sw2-Vlan-interface10]ip add 10.1.3.1 24
[sw2-Vlan-interface10]int v 30
[sw2-Vlan-interface30]ip add 10.1.4.2 24
```

这时 10.1.3.0 网段和 10.1.4.1，能够互通。请测试连通性并查看路由表。

```
[sw2]dis ip rou
  Routing Table: public net
Destination/Mask    Protocol    Pre   Cost    Nexthop      Interface
10.1.3.0/24         DIRECT      0     0       10.1.3.1     Vlan-interface10
10.1.3.1/32         DIRECT      0     0       127.0.0.1    InLoopBack0
10.1.4.0/24         DIRECT      0     0       10.1.4.2     Vlan-interface30
10.1.4.2/32         DIRECT      0     0       127.0.0.1    InLoopBack0
127.0.0.0/8         DIRECT      0     0       127.0.0.1    InLoopBack0
127.0.0.1/32        DIRECT      0     0       127.0.0.1    InLoopBack0
```

思考题：为什么 PC1 ping 不通 10.1.4.2？

(4) 配置交换机间的静态路由。

SW1 的配置：

```
[sw1]ip route-static 10.1.3.0 24 10.1.4.2
```

SW2 的配置:

    [sw2]ip route-static 10.1.1.0 24 10.1.4.1
    [sw2]ip route-static 10.1.2.0 24 10.1.4.1

配置完成后，全网络达到互通，可以看到路由表信息：

[sw1]dis ip routing-table
Routing Table: public net

Destination/Mask	Protocol	Pre	Cost	Nexthop	Interface
10.1.1.0/24	DIRECT	0	0	10.1.1.1	Vlan-interface10
10.1.1.1/32	DIRECT	0	0	127.0.0.1	InLoopBack0
10.1.2.0/24	DIRECT	0	0	10.1.2.1	Vlan-interface20
10.1.2.1/32	DIRECT	0	0	127.0.0.1	InLoopBack0
10.1.3.0/24	**STATIC**	**60**	**0**	**10.1.4.2**	**Vlan-interface30**
10.1.4.0/24	DIRECT	0	0	10.1.4.1	Vlan-interface30
10.1.4.1/32	DIRECT	0	0	127.0.0.1	InLoopBack0
127.0.0.0/8	DIRECT	0	0	127.0.0.1	InLoopBack0
127.0.0.1/32	DIRECT	0	0	127.0.0.1	InLoopBack0

[sw2]dis ip routing-table
Routing Table: public net

Destination/Mask	Protocol	Pre	Cost	Nexthop	Interface
10.1.1.0/24	**STATIC**	**60**	**0**	**10.1.4.1**	**Vlan-interface30**
10.1.2.0/24	**STATIC**	**60**	**0**	**10.1.4.1**	**Vlan-interface30**
10.1.3.0/24	DIRECT	0	0	10.1.3.1	Vlan-interface10
10.1.3.1/32	DIRECT	0	0	127.0.0.1	InLoopBack0
10.1.4.0/24	DIRECT	0	0	10.1.4.2	Vlan-interface30
10.1.4.2/32	DIRECT	0	0	127.0.0.1	InLoopBack0
127.0.0.0/8	DIRECT	0	0	127.0.0.1	InLoopBack0
127.0.0.1/32	DIRECT	0	0	127.0.0.1	InLoopBack0

（5）查看 ipfdb 表。

在三层交换机中，有关三层数据的转发主要是基于 ipfdb 表进行的。可以使用 "display ipfdb all" 命令来查看相关的 ipfdb 信息。

这里使用的 E328 没有这个命令，不过 ipfdb 表中的信息都是来源于 fib 和 arp 表，可以通过查看 fib 和 arp 表来了解三层数据的转发依据。

在 fib 表中，主要内容包括目的 IP 地址和下一跳 IP 地址，这些是和路由表一致的。

[sw1]dis fib
Flag:
  U:Usable   G:Gateway   H:Host   B:Blackhole   D:Dynamic   S:Static
  R:Reject   E:Equal cost multi-path   L:Generated by ARP or ESIS

Destination/Mask	Nexthop	Flag	TimeStamp	Interface
10.1.3.0/24	10.1.4.2	GSU	t[14039]	Vlan-interface30
10.1.4.1/32	127.0.0.1	GHU	t[13661]	InLoopBack0
10.1.2.1/32	127.0.0.1	GHU	t[13661]	InLoopBack0
10.1.4.0/24	10.1.4.1	U	t[13661]	Vlan-interface30

10.1.2.0/24	10.1.2.1	U	t[13661]	Vlan-interface20	
10.1.1.1/32	127.0.0.1	GHU	t[13261]	InLoopBack0	
10.1.1.0/24	10.1.1.1	U	t[13261]	Vlan-interface10	
127.0.0.1/32	127.0.0.1	GHU	t[26]	InLoopBack0	
127.0.0.0/8	127.0.0.1	U	t[26]	InLoopBack0	

在 arp 表中，主要是 IP 地址和 MAC 地址以及端口之间的对应关系。

[sw1]dis arp
　　　　　　Type: S-Static　　D-Dynamic

IP Address	MAC Address	VLAN ID	Port Name / AL ID	Aging	Type
10.1.4.2	000f-e254-8703	30	Ethernet1/0/24	20	D
10.1.1.2	001e-9004-0216	10	Ethernet1/0/4	16	D

实际上，将 fib 表和 arp 表的内容结合起来，就形成了 ipfdb 表，而三层交换机正是根据 ipfdb 表来实现三层数据转发的。

8.4　项目总结与提高

（1）写出主要项目实施规划、步骤与实训所得的主要结论。
（2）如果需要一部分 VLAN 可以通信，另一部分不能互相通信，应该如何做。
（3）请课余时间详细研究 ipfbd 表的构成。

项目九　VLAN 间 RIP 协议配置

9.1　项目提出

小王在某公司上班，业务部和财务部等多个部门分别在不同交换机的不同 VLAN 下，如果需要这些部门都可以通信，他该怎么办？

9.2　项目分析

1. 项目实训目的

掌握 RIP 协议在三层交换机上的配置。

2. 项目实现功能

利用 RIP 协议使跨交换机的 VLAN 间可以通信。

3. 项目主要应用的技术介绍

RIP 协议的全称是一种内部网关协议（IGP），是一种动态路由选择，用于一个自治系统（AS）内的路由信息的传递。RIP 协议是基于距离矢量算法（DistanceVectorAlgorithms）的，它使用"跳数"，即 metric 来衡量到达目标地址的路由距离。这种协议的路由器只关心自己周围的世界，只与自己相邻的路由器交换信息，范围限制在 15 跳(15 度)之内。RIP 应用于 OSI 网络七层模型的网络层。

RIP-1 被提出较早，其中有许多缺陷。为了改善 RIP-1 的不足，在 RFC1388 中提出了改进的 RIP-2，并在 RFC1723 和 RFC2453 中进行了修订。RIP-2 定义了一套有效的改进方案，新的 RIP-2 支持子网路由选择，支持 CIDR，支持组播，并提供了验证机制。

随着 OSPF 和 IS-IS 的出现，许多人认为 RIP 已经过时了。但事实上 RIP 也有它自己的优点。对于小型网络，RIP 就所占带宽而言开销小，易于配置、管理和实现，并且 RIP 还在大量使用中。但 RIP 也有明显的不足，即当有多个网络时会出现环路问题。为了解决环路问题，IETF 提出了分割范围方法，即路由器不可以通过它得知路由的接口去宣告路由。分割范围解决了两个路由器之间的路由环路问题，但不能防止 3 个或多个路由器形成路由环路。触发更新是解决环路问题的另一方法，它要求路由器在链路发生变化时立即传输它的路由表。这加速了网络的聚合，但容易产生广播泛滥。总之，环路问题的解决需要消耗一定的时间和带宽。若采用 RIP 协议，其网络内部所经过的链路数不能超过 15，这使得 RIP 协议不适于大型网络。

V1 和 V2 区别：
①RIPv1 是有类路由协议，RIPv2 是无类路由协议。
②RIPv1 不能支持 VLSM，RIPv2 可以支持 VLSM。
③RIPv1 没有认证的功能，RIPv2 可以支持认证，并且有明文和 MD5 两种认证。
④RIPv1 没有手工汇总的功能，RIPv2 可以在关闭自动汇总的前提下，进行手工汇总。

⑤RIPv1 是广播更新，RIPv2 是组播更新。

⑥RIPv1 对路由没有标记的功能，RIPv2 可以对路由打标记（tag），用于过滤和做策略。

⑦RIPv1 发送的 updata 最多可以携带 25 条路由条目，RIPv2 在有认证的情况下最多只能携带 24 条路由。

⑧RIPv1 发送的 updata 包里面没有 next-hop 属性，RIPv2 有 next-hop 属性，可以用于路由更新的重定。

9.3 项目实施

1. 项目拓扑图

VLAN 间 RIP 协议配置如图 9-1 所示。

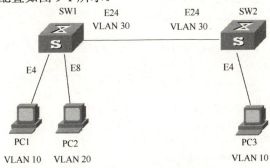

图 9-1　VLAN 间 RIP 协议配置

2. 项目实训环境准备

E328（2 台）、计算机（3 台）。

为了不受原来的配置影响，在实训之前先将所有的配置数据擦除后重新启动，命令为：

    <h3c>Reset saved-configuration
    <h3c>reboot

3. 项目主要实训步骤

按照拓扑图连接所有设备。检查设备的软件版本及配置信息，所有配置为初始状态。如果配置不符合要求，请在用户视图下擦除设备中的配置文件（reset saved-configuration），然后重启设备（reboot）以使系统采用缺省的配置参数进行初始化。

（1）初始配置。

配置表见表 9-1。

表 9-1　配置表

PC1		VLAN10	10.1.1.2/24	网关：10.1.1.1
PC2		VLAN20	10.1.2.2/24	网关：10.1.2.1
PC3		VLAN10	10.1.3.2/24	网关：10.1.3.1
SW1:VLAN10		SW1:E1～E6	10.1.1.1/24	
SW1:VLAN20		SW1:E7～E12	10.1.2.1/24	
SW1:VLAN30			10.1.4.1/24	
SW2: VLAN10		SW2:E1～E6	10.1.3.1/24	
SW2: VLAN30			10.1.4.2/24	

(2) 配置交换机 VLAN, TRUNK 端口。

SW1 的配置:

```
[H3C]sys sw1
[sw1]vlan 10
[sw1-vlan10]port e1/0/1 t e1/0/6
[sw1-vlan10]vlan 20
[sw1-vlan20]port e1/0/7 t e 1/0/12
[sw1-vlan20]vlan 30
[sw1-vlan30]quit
[sw1]int e1/0/24
[sw1-Ethernet1/0/24]port link-type trunk
[sw1-Ethernet1/0/24]port trunk permit vlan all
  Please wait........................... Done.
```

SW2 的配置:

```
[H3C]sys sw2
[sw2]vlan 10
[sw2-vlan10]port e1/0/1 t e1/0/6
[sw2]vlan 30
[sw2-vlan30]int e1/0/24
[sw2-Ethernet1/0/24]port link-type trunk
[sw2-Ethernet1/0/24]port trunk permit vlan all
  Please wait........................... Done.
```

(3) 配置 PC 和交换机的网段地址。

SW1 的配置:

```
[sw1]int Vlan-interface 10
[sw1-Vlan-interface10]ip add 10.1.1.1 24
[sw1-Vlan-interface10]int v 20
[sw1-Vlan-interface20]ip add 10.1.2.1 24
[sw1-Vlan-interface20]int v 30
[sw1-Vlan-interface30]ip add 10.1.4.1 24
```

这时 10.1.1.0 网段、10.1.2.0 网段和 10.1.4.1 能够互通。请测试连通性并查看路由表。

```
[sw1]dis ip rou
  Routing Table: public net
```

Destination/Mask	Protocol	Pre	Cost	Nexthop	Interface
10.1.1.0/24	DIRECT	0	0	10.1.1.1	Vlan-interface10
10.1.1.1/32	DIRECT	0	0	127.0.0.1	InLoopBack0
10.1.2.0/24	DIRECT	0	0	10.1.2.1	Vlan-interface20
10.1.2.1/32	DIRECT	0	0	127.0.0.1	InLoopBack0
10.1.4.0/24	DIRECT	0	0	10.1.4.1	Vlan-interface30
10.1.4.1/32	DIRECT	0	0	127.0.0.1	InLoopBack0
127.0.0.0/8	DIRECT	0	0	127.0.0.1	InLoopBack0
127.0.0.1/32	DIRECT	0	0	127.0.0.1	InLoopBack0

SW2 的配置：

[sw2]int v 10
[sw2-Vlan-interface10]ip add 10.1.3.1 24
[sw2-Vlan-interface10]int v 30
[sw2-Vlan-interface30]ip add 10.1.4.2 24

这时 10.1.3.0 网段和 10.1.4.1 能够互通。请测试连通性并查看路由表。

[sw2]dis ip rou
Routing Table: public net

Destination/Mask	Protocol	Pre	Cost	Nexthop	Interface
10.1.3.0/24	DIRECT	0	0	10.1.3.1	Vlan-interface10
10.1.3.1/32	DIRECT	0	0	127.0.0.1	InLoopBack0
10.1.4.0/24	DIRECT	0	0	10.1.4.2	Vlan-interface30
10.1.4.2/32	DIRECT	0	0	127.0.0.1	InLoopBack0
127.0.0.0/8	DIRECT	0	0	127.0.0.1	InLoopBack0
127.0.0.1/32	DIRECT	0	0	127.0.0.1	InLoopBack0

思考题：为什么 PC1 ping 不通 10.1.4.2？

（4）配置交换机间的 RIP 协议。

SW1 的配置：

[sw1]rip
[sw1-rip]network 10.1.1.0
[sw1-rip]network 10.1.2.0
[sw1-rip]network 10.1.4.0

SW2 的配置：

[sw2]rip
[sw2-rip]network 10.1.3.0
[sw2-rip]network 10.1.4.0

配置完成后，全网络达到互通，可以看到路由表信息：

[sw1]dis ip routing-table
Routing Table: public net

Destination/Mask	Protocol	Pre	Cost	Nexthop	Interface
10.1.1.0/24	DIRECT	0	0	10.1.1.1	Vlan-interface10
10.1.1.1/32	DIRECT	0	0	127.0.0.1	InLoopBack0
10.1.2.0/24	DIRECT	0	0	10.1.2.1	Vlan-interface20
10.1.2.1/32	DIRECT	0	0	127.0.0.1	InLoopBack0
10.1.3.0/24	**RIP**	**100**	**1**	**10.1.4.2**	**Vlan-interface30**
10.1.4.0/24	DIRECT	0	0	10.1.4.1	Vlan-interface30
10.1.4.1/32	DIRECT	0	0	127.0.0.1	InLoopBack0
127.0.0.0/8	DIRECT	0	0	127.0.0.1	InLoopBack0
127.0.0.1/32	DIRECT	0	0	127.0.0.1	InLoopBack0

[sw2]dis ip routing-table
Routing Table: public net

Destination/Mask	Protocol	Pre	Cost	Nexthop	Interface
10.1.1.0/24	**RIP**	**100**	**1**	**10.1.4.1**	**Vlan-interface30**
10.1.2.0/24	**RIP**	**100**	**1**	**10.1.4.1**	**Vlan-interface30**
10.1.3.0/24	DIRECT	0	0	10.1.3.1	Vlan-interface10
10.1.3.1/32	DIRECT	0	0	127.0.0.1	InLoopBack0
10.1.4.0/24	DIRECT	0	0	10.1.4.2	Vlan-interface30
10.1.4.2/32	DIRECT	0	0	127.0.0.1	InLoopBack0
127.0.0.0/8	DIRECT	0	0	127.0.0.1	InLoopBack0
127.0.0.1/32	DIRECT	0	0	127.0.0.1	InLoopBack0

（5）查看 RIP 信息。

  [sw1]dis rip
   RIP is running
    Checkzero is on   Default cost : 1
    Summary is on   Preference : 100
    Traffic-share-across-interface is off
    Period update timer : 30
    Timeout timer : 180
    Garbage-collection timer : 120
    No peer router
    Network :
    10.0.0.0

  [sw2]dis rip
   RIP is running
    Checkzero is on   Default cost : 1
    Summary is on   Preference : 100
    Traffic-share-across-interface is off
    Period update timer : 30
    Timeout timer : 180
    Garbage-collection timer : 120
    No peer router
    Network :
    10.0.0.0

打开 RIP 协议报文的调试开关。

  [sw2]quit
  <sw2>debugging rip packet
   Rip packet debugging is on

将调试信息输出到终端。

  <sw2>terminal debugging
  % Current terminal debugging is on
  *0.30972203 sw2 RM/7/RTDBG:- 1 -
  RIP: send from 10.1.4.2(Vlan-interface30) to 255.255.255.255
    Packet:vers 1, cmd Response, length 24
    dest 10.1.3.0 , metric 1, tag 0

```
*0.31018868 sw2 RM/7/RTDBG:- 1 -
ignoring RIP Response
        1 packet from 10.1.1.1+520 - can not find interface for source address
*0.31019025 sw2 RM/7/RTDBG:- 1 -
RIP: Receive Response from 10.1.4.1+520 via lan-interface30(255.255.255.255)
    Packet:vers 1, cmd Response, length 44
        dest 10.1.2.0    , metric 1, tag 0
        dest 10.1.1.0    , metric 1, tag 0
```

从上面输出信息可以看到，SW2 从路由接口 VLAN30(10.1.4.2)给广播地址 255.255.255.255（代表所有运行 RIPv1 协议的路由器）发送 RIPv1 路由更新报文，携带一条为 10.1.3.0 / 24 的路由；同时从 SW1 的路由接口 10.1.4.1 收到 RIPv1 路由更新报文，包括 2 条路由：10.1.1.0/24,10.1.2.0/24。

运行 RIP 协议报文时，接口在缺省情况下接收和发送 rip-1 报文。而当指定接口 RIP 版本为 rip-2 时，缺省使用组播形式传送报文。

```
[sw2]int v 30
[sw2-Vlan-interface30]rip version 2 multicast
[sw2-Vlan-interface30]quit
```

调试信息中，可以看到，从 SW2 发出的 RIP 信息，目的地址是 224.0.0.9（代表所有运行 RIPv2 协议的路由器）

```
*0.31872217 sw2 RM/7/RTDBG:- 1 -
 RIP: send from 10.1.4.2(Vlan-interface30) to 224.0.0.9
    Packet:vers 2, cmd Response, length 24
        dest 10.1.3.0    mask 255.255.255.0, router 0.0.0.0    , metric 1, tag 0


<sw2>u t d
% Current terminal debugging is off
```

9.4 项目总结与提高

（1）写出主要项目实施规划、步骤与实训所得的主要结论。
（2）如果需要一部分 VLAN 可以通信，另一部分不能互相通信，应该如何做。

项目十 高级 STP 配置

10.1 项目提出

小王在某公司上班,虽然他在网络设备上配置了生成树协议防止环路存在,但搞不明白哪里是根,哪里会自动断开,应该如何理解生成树协议呢?

10.2 项目分析

1. 项目实训目的

理解 STP 的基本原理和生成树的生成过程。

验证 STP 端口状态的切换。

2. 项目实现功能

通过实训,理解生成树的原理。

3. 项目主要应用的技术介绍

STP(生成树协议)、RSTP(快速生成树协议)、MSTP(多生成树协议),这三个协议都是二层交换网络中为了防止环路和实现链路冗余而设计的。

(1)STP(802.1d)。

STP 协议生来就是为了冗余而存在的,单纯树型的网络无法提供足够的可靠性,由此我们引入了额外的链路,这才出现了环路这样的问题。但标准的 802.1d STP 协议并不能实现真正的冗余与负载分担。

STP 内部只有一棵 STP tree,因此必然有一条链路要被 blocking,不会转发数据,只有另外一条链路出现问题时,这条被 blocking 的链路才会接替之前链路所承担的职责,做数据的转发。无论怎样,总会有一条链路处于不被使用的状态,冗余是有了,但是负载分担没有做到。

随着网络的发展,人们发现传统的 STP 协议无法满足主备快速切换的需求,因为 STP 协议将端口定义了 5 种状态,分别为:blocking、listening、learning、forwarding、disabling,想要从 blocking 切换至 forwarding 状态,必需要经过 50 秒的周期,这 50 秒我们只能被动地去等待。20 秒的 blocking 状态下,如果没有检测到邻居发来的 BPDU 包,则进入 listening,这时要做的是选举 Root Bridge、Designate Port、Root Port,15 秒后,进入 learning,learning 状态下可以学习 MAC 地址,为最后的 forwarding 做准备,同样是 15 秒,最后到达转发状态。这样的延时在现代网络环境下是让人极为难以忍受的。

(2)RSTP(802.1w)。

RSTP 的出现解决了延时的问题,它的收敛速度很快。RSTP 在 STP 基础上额外定义了两种 port role(端口角色),分别是 alternate 与 backup。另外重新规定了 port state(端口状态),分别为 discarding、Learning、Forwarding。

STP 的一大失败之处在于混淆了 port role 与 port state 两种概念,在 RSTP 上,这样的问题不再存在了，port state 与 port role 无关了。alternate port 责任是为另一台交换机上的链路做备份，而 backup port 是为本交换机上的端口做备份。

RSTP 最重要的变化在于对 BPDU 中 type 字段的利用上，除了之前 STP 使用的两个位，在另外 6 个位中实现了很多的功能，包括不再需要去等待 50 秒的时间完成主备切换，直接利用 proposal 与 agreement 协商即可，这样大大缩短了收敛时间。

RSTP 还定义了两个新的概念：edge port 与 link type，如果是 edge port，表明下面接的只能是主机，环路的存在是不可能的，所以可以直接将其从 discarding 切换到 forwarding 状态，类似于 STP 中的 port fast 技术。而 link type 定义了这条链路是 point-to-point 的还是 shared。如果有 pt-pt 环境下，就可以做快速的切换。

（3）MSTP（802.1s）。

STP 和 RSTP 都采用了一棵 STP tree，负载分担不可实现，这也正是 MSTP（802.1s）产生的原因。

MSTP 可以将多个 VLAN 的生成树映射为一个实例，即 vlan map to a instance，实际上不需要那么多的生成树，只需要按照冗余链路的条数来得出生成树的数量。

如果只有两条链路，并且有 1~1000 个 VLAN，可以将 1~500 定义为 instance 1，将 501~1000 定义到 instance 2。只生成两棵树 1 和 2，实现了冗余与负载分担。

MSTP 是基于 RSTP 的，没有 RSTP，MSTP 是无法运行的。

（4）STP、RSTP、MSTP 的对比分析。

①STP 不能使端口状态快速迁移，即使是在点对点链路或边缘端口，也必须等待 2 倍的 Forward delay 的时间延迟，端口才能迁移到转发状态。

②RSTP 可以快速收敛，但是和 STP 一样存在以下缺陷：局域网内所有网桥共享一棵生成树，不能按 VLAN 阻塞冗余链路，所有 VLAN 的报文都沿着一棵生成树进行转发。

③MSTP 将环路网络修剪成为一个无环的树型网络，避免报文在环路网络中的增生和无限循环，同时还提供了数据转发的多个冗余路径，在数据转发过程中实现 VLAN 数据的负载均衡。

④MSTP 兼容 STP 和 RSTP，并且可以弥补 STP 和 RSTP 的缺陷。它既可以快速收敛，也能使不同 VLAN 的流量沿各自的路径分发，从而为冗余链路提供了更好的负载分担。

10.3 项目实施

1. 项目拓扑图

高级 STP 配置如图 10-1 所示。

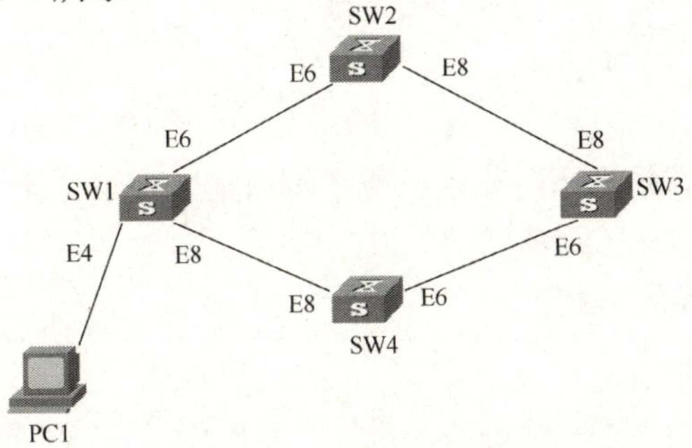

图 10-1 高级 STP 配置

2. 项目实训环境准备

E126（4台）、计算机（1台）。

为了不受原来的配置影响，在实训之前先将所有的配置数据擦除后重新启动，命令为：

<h3c>Reset saved-configuration
<h3c>reboot

3. 项目主要实训步骤

按照拓扑图连接所有设备。检查设备的软件版本及配置信息，所有配置为初始状态。如果配置不符合要求，请在用户视图下擦除设备中的配置文件（reset saved-configuration），然后重启设备（reboot）以使系统采用缺省的配置参数进行初始化。

（1）初始配置。

首先，查看4个交换机的MAC地址。

SW1 的 MAC 地址：

    [sw1]dis int e1/0/8
    Ethernet1/0/8 current state : UP
    IP Sending Frames' Format is PKTFMT_ETHNT_2, Hardware address is **000f-e25c-4106**

SW2 的 MAC 地址：

    [sw2]dis int e1/0/8
    Ethernet1/0/8 current state : UP
    IP Sending Frames' Format is PKTFMT_ETHNT_2, Hardware address is **000f-e25c-40f7**

SW3 的 MAC 地址：

    [sw3]dis int e1/0/8
    Ethernet1/0/8 current state : UP
    IP Sending Frames' Format is PKTFMT_ETHNT_2, Hardware address is **000f-e25c-4116**

SW4 的 MAC 地址：

    [sw4]dis int e1/0/8
    Ethernet1/0/8 current state : UP
    IP Sending Frames' Format is PKTFMT_ETHNT_2, Hardware address is **000f-e25c-40f6**

在H3C交换机上启动STP协议：

    [sw1]stp enable
    [sw2]stp enable
    [sw3]stp enable
    [sw4]stp enable

全网配置STP后，默认情况下，交换机每个端口都启用了STP协议。可以看到交换机指示灯不再快速闪烁，已经建立了无环路的转发生成树。

（2）生成树协议算法实现过程。

①初始状态。

每台交换机的各个端口在初始时生成以自己为根的配置消息，根路径开销为0，指定交换机ID为自身交换机ID，指定端口为本端口。

配置消息格式：{根 ID，根路径开销，指定交换机 ID，指定端口 ID}

SW1：

端口 E6 配置消息：

  {32768.000f-e25c-4106，0，32768.000f-e25c-4106，e1/0/6}

端口 E8 配置消息：

  {32768.000f-e25c-4106，0，32768.000f-e25c-4106，e1/0/8}

SW2：

端口 E6 配置消息：

  {32768.000f-e25c-40f7，0，32768.000f-e25c-40f7，e1/0/6}

端口 E8 配置消息：

  {32768.000f-e25c-40f7，0，32768.000f-e25c-40f7，e1/0/8}

SW3：

端口 E6 配置消息：

  {32768.000f-e25c-4116，0，32768.000f-e25c-4116，e1/0/6}

端口 E8 配置消息：

  {32768.000f-e25c-4116，0，32768.000f-e25c-4116，e1/0/8}

SW4：

端口 E6 配置消息：

  {32768.000f-e25c-40f6，0，32768.000f-e25c-40f6，e1/0/6}

端口 E8 配置消息：

  {32768.000f-e25c-40f6，0，32768.000f-e25c-40f6，e1/0/8}

②选出最优配置消息，确定根交换机。

当某个端口收到比自身的配置消息优先级低的配置消息时，交换机会将接收到的配置消息丢弃，对该端口的配置消息不做任何处理。

当端口收到比自身优先级高的配置消息时，用接收到的配置消息中的内容替换该端口的配置消息。

然后交换机将该端口的配置消息和交换机上的其他端口进行比较，选出最优的配置消息。

配置消息的比较原则是：

- 树根 ID 较小的配置消息优先级高；
- 若树根 ID 相同，则比较根路径开销（用配置消息中的根路径开销加上本端口对应的路径开销之和，则较小的配置消息优先级较高）；
- 若根路径开销也相同，则依次比较指定交换机 ID、指定端口 ID、接收该配置消息的端口 ID 等。

根据原则，先比较交换机 ID，由于交换机的 ID 有交换机的优先级（缺省：32768）和交换机 MAC 地址共同组成。起初优先级都是缺省值。所以，谁的 MAC 地址最小，谁就是根。

很显然，根交换机是 SW4。可以用命令查看配置后的 STP 信息：

```
[sw1]dis stp
-------[CIST Global Info][Mode MSTP]-------
CIST Bridge            :32768.000f-e25c-4106
Bridge Times           :Hello 2s MaxAge 20s FwDly 15s MaxHop 20
CIST Root/ERPC         :32768.000f-e25c-40f6 / 200
CIST RegRoot/IRPC      :32768.000f-e25c-4106 / 0
CIST RootPortId        :128.8
BPDU-Protection        :disabled
TC-Protection          :enabled / Threshold=6
Bridge Config
Digest Snooping        :disabled
TC or TCN received     :13
Time since last TC     :0 days 0h:10m:7s

[sw2]dis stp
-------[CIST Global Info][Mode MSTP]-------
CIST Bridge            :32768.000f-e25c-40f7
Bridge Times           :Hello 2s MaxAge 20s FwDly 15s MaxHop 20
CIST Root/ERPC         :32768.000f-e25c-40f6 / 400
CIST RegRoot/IRPC      :32768.000f-e25c-40f7 / 0
CIST RootPortId        :128.6
BPDU-Protection        :disabled
TC-Protection          :enabled / Threshold=6
Bridge Config
Digest Snooping        :disabled
TC or TCN received     :25
Time since last TC     :0 days 0h:11m:24s

[sw3]dis stp
-------[CIST Global Info][Mode MSTP]-------
CIST Bridge            :32768.000f-e25c-4116
Bridge Times           :Hello 2s MaxAge 20s FwDly 15s MaxHop 20
CIST Root/ERPC         :32768.000f-e25c-40f6 / 200
CIST RegRoot/IRPC      :32768.000f-e25c-4116 / 0
CIST RootPortId        :128.6
BPDU-Protection        :disabled
TC-Protection          :enabled / Threshold=6
Bridge Config
Digest Snooping        :disabled
TC or TCN received     :6
Time since last TC     :0 days 0h:15m:25s
[sw4]dis stp
-------[CIST Global Info][Mode MSTP]-------
CIST Bridge            :32768.000f-e25c-40f6
Bridge Times           :Hello 2s MaxAge 20s FwDly 15s MaxHop 20
CIST Root/ERPC         :32768.000f-e25c-40f6 / 0
```
注释：CIST 根/CIST 外部根路径开销

CIST RegRoot/IRPC   :32768.000f-e25c-40f6 / 0
注释：CIST 区域根/ CIST 内部根路径开销
CIST RootPortId   :0.0
BPDU-Protection   :disabled
TC-Protection   :enabled / Threshold=6
Bridge Config
Digest Snooping   :disabled
TC or TCN received   :7
Time since last TC   :0 days 0h:12m:31s

从上 4 台交换机所示，根交换机是 32768.000f-e25c-40f6，也就是 SW4。

③确定根端口，并阻塞冗余链路，然后更新指定端口的配置消息。

交换机接收最优配置消息的那个端口定为根端口，端口配置消息不作改变；

其他端口中，如果某端口的配置消息在"选出最优配置消息"的过程中更新过，则交换机将此端口阻塞，端口配置消息不变，此端口将不再转发数据，并且只接收但不发送配置消息；

如果某端口的配置消息在"选出最优配置消息"的过程中没有更新，则交换机就将其定为指定端口，配置消息要作如下改变：树根 ID 替换为根端口的配置消息的树根 ID，根路径开销替换为根端口的配置消息的根路径开销加上根端口对应的路径开销，指定交换机 ID 替换为自身交换机的 ID，指定端口 ID 替换为自身端口 ID。

如本例中：

SW1：

端口 6 {32768.000f-e25c-4106，0，32768.000f-e25c-4106，e1/0/6}

收到 SW2 的配置消息{32768.**000f-e25c-40f7**，0，32768.**000f-e25c-40f7**，e1/0/6}，

更新：{32768.000f-e25c-40f7，200，32768.000f-e25c-40f7，e1/0/6}

端口 8 {32768.000f-e25c-4106，0，32768.000f-e25c-4106，e1/0/8}

收到 SW4 的配置消息{32768.**000f-e25c-40f6**，0，32768.**000f-e25c-40f6**，e1/0/6}，

更新：{32768.000f-e25c-40f6，200，32768.000f-e25c-40f6，e1/0/8}

SW1 对各个端口的配置消息进行比较，选出端口 E8 的配置消息为最优配置消息，然后将端口 E8 定为根端口，其他端口的配置消息更新。

端口 E8（不变）：{32768.000f-e25c-40f6，200，32768.000f-e25c-40f6，e1/0/8}

端口 E6（更新）：{32768.000f-e25c-40f6，200，32768.000f-e25c-4106，e1/0/6}

然后从各个端口周期性向外发送。

SW4

配置消息优先级最高{32768.**000f-e25c-40f6**，0，32768.**000f-e25c-40f6**，e1/0/6}，所以不变，向外发送。

端口 E6：{32768.**000f-e25c-40f6**，0，32768.**000f-e25c-40f6**，e1/0/6}

端口 E8：{32768.**000f-e25c-40f6**，0，32768.**000f-e25c-40f6**，e1/0/8}

SW3

端口 E6：{32768.**000f-e25c-4116**，0，32768.**000f-e25c-4116**，e1/0/6}

收到：SW4 的{32768.**000f-e25c-40f6**，0，32768.**000f-e25c-40f6**，e1/0/6}

更新为：{32768.**000f-e25c-40f6**，200，32768.**000f-e25c-40f6**，e1/0/6}
端口 E8：{32768.**000f-e25c-4116**，0，32768.**000f-e25c-4116**，e1/0/8}
收到：SW2 的{32768.**000f-e25c-40f7**，0，32768.**000f-e25c-40f7**，e1/0/8}
更新为：{32768.**000f-e25c-40f7**，0，32768.**000f-e25c-40f7**，e1/0/8}
比较后：
端口 E6（不变）：{32768.**000f-e25c-40f6**，200，32768.**000f-e25c-40f6**，e1/0/6}
端口 E8（更新）：{32768.**000f-e25c-40f6**，200，32768.**000f-e25c-4116**，e1/0/8}

SW2
端口 E6：{32768.**000f-e25c-40f7**，0，32768.**000f-e25c-40f7**，e1/0/6}
收到：SW1 的{32768. 000f-e25c-40f6，200，32768. 000f-e25c-4106，e1/0/6}
更新为：{32768. 000f-e25c-40f6，400，32768. 000f-e25c-4106，e1/0/6}
端口 E8：{32768.**000f-e25c-40f7**，0，32768.**000f-e25c-40f7**，e1/0/8}
收到：SW3 {32768.**000f-e25c-40f6**，200，32768.**000f-e25c-4116**，e1/0/8}
更新为：{32768.**000f-e25c-40f6**，400，32768.**000f-e25c-4116**，e1/0/8}
比较后：
端口 E8（阻断，不变）：{32768.**000f-e25c-40f6**，400，32768.**000f-e25c-4116**，e1/0/8}
端口 E6（不变）：{32768. 000f-e25c-40f6，400，32768. 000f-e25c-4106，e1/0/6}

（3）查看端口 STP 信息。

可以通过查看端口 STP 信息，得到验证

```
[sw1]display stp brief
 MSTID        Port              Role    STP State     Protection
   0          Ethernet1/0/6     DESI    FORWARDING    NONE
   0          Ethernet1/0/8     ROOT    FORWARDING    NONE
[sw1]dis stp int e1/0/6 e1/0/8
----[CIST][Port6(Ethernet1/0/6)][FORWARDING]----
 Port Protocol          :enabled
 Port Role              :CIST Designated Port
 Port Priority          :128
 Port Cost(Legacy)      :Config=auto / Active=200
 Desg. Bridge/Port      :32768.000f-e25c-4106 / 128.6
 Port Edged             :Config=disabled / Active=disabled
 Point-to-point         :Config=auto / Active=true
 Transmit Limit         :10 packets/hello-time
 Protection Type        :None
 MSTP BPDU format       :Config=legacy
 Port Config
 Digest Snooping        :disabled
 Num of Vlans Mapped :1
 PortTimes              :Hello 2s MaxAge 20s FwDly 15s MsgAge 1s RemHop 20
 BPDU Sent              :1350
         TCN: 0, Config: 0, RST: 0, MST: 1350
 BPDU Received          :35
         TCN: 0, Config: 0, RST: 0, MST: 35
```

项目十 高级STP配置

```
----[CIST][Port8(Ethernet1/0/8)][FORWARDING]----
 Port Protocol           :enabled
 Port Role               :CIST Root Port
Port Priority            :128
 Port Cost(Legacy)       :Config=auto / Active=200
 Desg. Bridge/Port       :32768.000f-e25c-40f6 / 128.8
 Port Edged              :Config=disabled / Active=disabled
 Point-to-point          :Config=auto / Active=true
 Transmit Limit          :10 packets/hello-time
 Protection Type         :None
 MSTP BPDU format        :Config=legacy
 Port Config
 Digest Snooping         :disabled
 Num of Vlans Mapped :1
 PortTimes               :Hello 2s MaxAge 20s FwDly 15s MsgAge 0s RemHop 20
 BPDU Sent               :7
         TCN: 0, Config: 0, RST: 0, MST: 7
 BPDU Received           :1328
         TCN: 0, Config: 0, RST: 0, MST: 1328

[sw2]display stp brief
  MSTID       Port              Role    STP State      Protection
    0         Ethernet1/0/6     ROOT    FORWARDING     NONE
    0         Ethernet1/0/8     ALTE    DISCARDING     NONE
[sw2]dis stp int e1/0/6 e1/0/8
----[CIST][Port6(Ethernet1/0/6)][FORWARDING]----
 Port Protocol           :enabled
 Port Role               :CIST Root Port
 Port Priority           :128
 Port Cost(Legacy)       :Config=auto / Active=200
 Desg. Bridge/Port       :32768.000f-e25c-4106 / 128.6
 Port Edged              :Config=disabled / Active=disabled
 Point-to-point          :Config=auto / Active=true
 Transmit Limit          :10 packets/hello-time
 Protection Type         :None
 MSTP BPDU format        :Config=legacy
 Port Config
 Digest Snooping         :disabled
 Num of Vlans Mapped :1
 PortTimes               :Hello 2s MaxAge 20s FwDly 15s MsgAge 1s RemHop 20
 BPDU Sent               :35
         TCN: 0, Config: 0, RST: 0, MST: 35
 BPDU Received           :1926
         TCN: 0, Config: 0, RST: 0, MST: 1926
----[CIST][Port8(Ethernet1/0/8)][DISCARDING]----
 Port Protocol           :enabled
 Port Role               :CIST Alternate Port
```

Port Priority   :128
Port Cost(Legacy)   :Config=auto / Active=200
Desg. Bridge/Port   :**32768.000f-e25c-4116** / 128.8
Port Edged   :Config=disabled / Active=disabled
Point-to-point   :Config=auto / Active=true
Transmit Limit   :10 packets/hello-time
Protection Type   :None
MSTP BPDU format   :Config=legacy
Port Config
Digest Snooping   :disabled
Num of Vlans Mapped :1
PortTimes   :Hello 2s MaxAge 20s FwDly 15s MsgAge 1s RemHop 20
BPDU Sent   :3
      TCN: 0, Config: 0, RST: 0, MST: 3
BPDU Received   :1821
      TCN: 0, Config: 0, RST: 0, MST: 1821

[sw3]display stp brief

MSTID	Port	Role	STP State	Protection
0	Ethernet1/0/6	ROOT	FORWARDING	NONE
0	Ethernet1/0/8	DESI	FORWARDING	NONE

[sw3]dis stp int e1/0/6 e1/0/8
----[CIST][Port6(Ethernet1/0/6)][FORWARDING]----
Port Protocol   :enabled
Port Role   :CIST Root Port
Port Priority   :128
Port Cost(Legacy)   :Config=auto / Active=200
Desg. Bridge/Port   :**32768.000f-e25c-40f6** / 128.6
Port Edged   :Config=disabled / Active=disabled
Point-to-point   :Config=auto / Active=true
Transmit Limit   :10 packets/hello-time
Protection Type   :None
MSTP BPDU format   :Config=legacy
Port Config
Digest Snooping   :disabled
Num of Vlans Mapped :1
PortTimes   :Hello 2s MaxAge 20s FwDly 15s MsgAge 0s RemHop 20
BPDU Sent   :18
      TCN: 0, Config: 0, RST: 0, MST: 18
BPDU Received   :2000
      TCN: 0, Config: 0, RST: 0, MST: 2000
----[CIST][Port8(Ethernet1/0/8)][FORWARDING]----
Port Protocol   :enabled
Port Role   :CIST Designated Port
Port Priority   :128
Port Cost(Legacy)   :Config=auto / Active=200
Desg. Bridge/Port   :**32768.000f-e25c-4116** / 128.8

Port Edged   :Config=disabled / Active=disabled
Point-to-point   :Config=auto / Active=true
Transmit Limit   :10 packets/hello-time
Protection Type   :None
MSTP BPDU format   :Config=legacy
Port Config
Digest Snooping   :disabled
Num of Vlans Mapped :1
PortTimes   :Hello 2s MaxAge 20s FwDly 15s MsgAge 1s RemHop 20
BPDU Sent   :1903
        TCN: 0, Config: 0, RST: 0, MST: 1903
BPDU Received   :3
        TCN: 0, Config: 0, RST: 0, MST: 3

[sw4]display stp brief
 MSTID   Port   Role   STP State   Protection
   0   Ethernet1/0/6   DESI   FORWARDING   NONE
   0   Ethernet1/0/8   DESI   FORWARDING   NONE
[sw4]dis stp int e1/0/6 e1/0/8
----[CIST][Port6(Ethernet1/0/6)][FORWARDING]----
Port Protocol   :enabled
Port Role   :CIST Designated Port
Port Priority   :128
Port Cost(Legacy)   :Config=auto / Active=200
Desg. Bridge/Port   :**32768.000f-e25c-40f6** / 128.6
Port Edged   :Config=disabled / Active=disabled
Point-to-point   :Config=auto / Active=true
Transmit Limit   :10 packets/hello-time
Protection Type   :None
MSTP BPDU format   :Config=legacy
Port Config
Digest Snooping   :disabled
Num of Vlans Mapped :1
PortTimes   :Hello 2s MaxAge 20s FwDly 15s MsgAge 0s RemHop 20
BPDU Sent   :2043
        TCN: 0, Config: 0, RST: 0, MST: 2043
BPDU Received   :7
        TCN: 0, Config: 0, RST: 0, MST: 7

----[CIST][Port8(Ethernet1/0/8)][FORWARDING]----
Port Protocol   :enabled
Port Role   :CIST Designated Port
Port Priority   :128
Port Cost(Legacy)   :Config=auto / Active=200
Desg. Bridge/Port   :**32768.000f-e25c-40f6** / 128.8
Port Edged   :Config=disabled / Active=disabled
Point-to-point   :Config=auto / Active=true

```
Transmit Limit          :10 packets/hello-time
Protection Type         :None
MSTP BPDU format        :Config=legacy
Port Config
Digest Snooping         :disabled
Num of Vlans Mapped :1
PortTimes               :Hello 2s MaxAge 20s FwDly 15s MsgAge 0s RemHop 20
BPDU Sent               :2034
    TCN: 0, Config: 0, RST: 0, MST: 2034
BPDU Received           :4
    TCN: 0, Config: 0, RST: 0, MST: 4
```

最后生成树被确定下来，树根为 SW4，树形如图 10-2 所示。

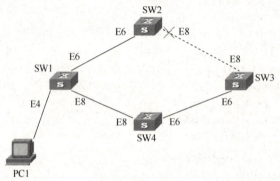

图 10-2　生成树

（4）端口状态切换。

生成树稳定状态下，SW2 的 E6 口处于 forwarding 状态，正常接收 BPDU 消息，同时能收发数据包。E8 口处于 discarding 状态，能正常接收 BPDU 消息，但不能收发数据包。

```
<sw2>debugging stp packet
<sw2>te
<sw2>terminal d
<sw2>terminal debugging
% Current terminal debuggin
*0.6785436 sw2 MSTP/8/PKT:- 1 -
Port6(Ethernet1/0/6) Rcvd Packet(Length: 103)
ProtocolVersionID: 03
BPDUType            : 02
Instance(Flags)     : 0(6c)

*0.6785754 sw2 MSTP/8/PKT:- 1 -
Port8(Ethernet1/0/8) Rcvd Packet(Length: 103)
ProtocolVersionID: 03
BPDUType            : 02
Instance(Flags)     : 0(6c)
```

现在，断开 SW1 和 SW2 之间的连线，这样 SW2 的 E6 口会 DOWN 下来，E8 口的状态会从 discarding 状态转换到 forwarding 状态，承担起数据收发工作。

```
<sw2>sys
System View: return to User View with Ctrl+Z.
[sw2]dis stp b
 MSTID        Port                 Role      STP State      Protection
   0          Ethernet1/0/8        ROOT      FORWARDING     NONE
```

因为 E6 口已经 DOWN 掉,所以看不见。

【参考资料】

1. MSTP 简介

STP(Spanning Tree Protocol,生成树协议)不能使端口状态快速迁移,即使是在点对点链路或边缘端口,也必须等待 2 倍的 Forward delay 的时间延迟,端口才能迁移到转发状态。

RSTP(Rapid Spanning Tree Protocol,快速生成树协议)可以快速收敛,但是和 STP 一样存在以下缺陷:局域网内所有网桥共享一棵生成树,不能按 VLAN 阻塞冗余链路,所有 VLAN 的报文都沿着一棵生成树进行转发。

MSTP(Multiple Spanning Tree Protocol,多生成树协议)将环路网络修剪成为一个无环的树型网络,避免报文在环路网络中的增生和无限循环,同时还提供了数据转发的多个冗余路径,在数据转发过程中实现 VLAN 数据的负载均衡。

MSTP 兼容 STP 和 RSTP,并且可以弥补 STP 和 RSTP 的缺陷。它既可以快速收敛,也能使不同 VLAN 的流量沿各自的路径分发,从而为冗余链路提供了更好的负载分担机制。

2. MSTP 的基本概念

MSTP 的基本概念示意图如图 10-3 所示。

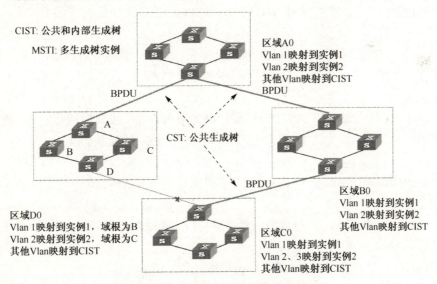

图 10-3 MSTP 的基本概念示意图

MST 域

MST 域(Multiple Spanning Tree Regions,多生成树域)是由交换网络中的多台交换机以及它们之间的网段构成的。这些交换机都启动了 MSTP、具有相同的域名、相同的 VLAN 到生成树映射配置和相同的 MSTP 修订级别配置,并且物理上有链路连通。

一个交换网络可以存在多个 MST 域。用户可以通过 MSTP 配置命令把多台交换机划分在同一个 MST 域内。例如图 10-3 中的区域 A0，域内所有交换机都有相同的 MST 域配置：域名相同，VLAN 与生成树的映射关系相同（VLAN1 映射到生成树实例 1，VLAN2 映射到生成树实例 2，其余 VLAN 映射到 CIST），MSTP 修订级别相同（此配置在图中没有体现）。

MSTI

MSTI（Multiple Spanning Tree Instance，多生成树实例）是指 MST 域内的生成树。

一个 MST 域内可以通过 MSTP 生成多棵生成树，各棵生成树之间彼此独立。每个域内可以存在多棵生成树，每棵生成树和相应的 VLAN 对应。这些生成树都被称为 MSTI。

VLAN 映射表

VLAN 映射表是 MST 域的一个属性，用来描述 VLAN 和 MSTI 的映射关系。区域 A0 的 VLAN 映射表就是：VLAN 1 映射到生成树实例 1，VLAN 2 映射到生成树实例 2，其余 VLAN 映射到 CIST。MSTP 就是根据 VLAN 映射表来实现负载分担的。

IST

IST（Internal Spanning Tree，内部生成树）是 MST 域内的一棵生成树。

IST 和 CST（Common Spanning Tree，公共生成树）共同构成整个交换机网络的生成树 CIST（Common and Internal Spanning Tree，公共和内部生成树）。IST 是 CIST 在 MST 域内的片段，是一个特殊的多生成树实例。CIST 在每个 MST 域内都有一个片段，这个片段就是各个域内的 IST。

CST

CST 是连接交换网络内所有 MST 域的单生成树。如果把每个 MST 域看作是一个"交换机"，CST 就是这些"交换机"通过 STP 协议、RSTP 协议计算生成的一棵生成树。

CIST

CIST 是连接一个交换网络内所有交换机的单生成树，由 IST 和 CST 共同构成。每个 MST 域内的 IST 加上 MST 域间的 CST 就构成整个网络的 CIST。

域根

域根是指 MST 域内 IST 和 MSTI 的树根。MST 域内各棵生成树的拓扑不同，域根也可能不同。区域 D0 中，生成树实例 1 的域根为交换机 B，生成树的域根为交换机 C。

总根

总根（Common Root Bridge）是指 CIST 的树根。总根为区域 A0 内的某台交换机。

端口角色

在 MSTP 的计算过程中，端口角色主要有根端口、指定端口、Master 端口、域边缘端口、Alternate 端口和 Backup 端口。

- 根端口是负责向树根方向转发数据的端口。
- 指定端口是负责向下游网段或交换机转发数据的端口。
- Master 端口是连接 MST 域到总根的端口，位于整个域到总根的最短路径上。
- 域边缘端口是连接不同 MST 域、MST 域和运行 STP 的区域、MST 域和运行 RSTP 的区域的端口，位于 MST 域的边缘。
- Alternate 端口是 Master 端口的备份端口，如果 Master 端口被阻塞后，Alternate 端口将成为新的 Master 端口。

- 当同一台交换机的两个端口互相连接时就存在一个环路，此时交换机会将其中一个端口阻塞，Backup 端口是被阻塞的那个端口。

端口状态

MSTP 中，根据端口是否转发用户流量、接收/发送 BPDU 报文，端口状态划分为三种：
- Forwarding 状态：既转发用户流量又接收/发送 BPDU 报文；
- Learning 状态：不转发用户流量，只接收/发送 BPDU 报文；
- Discarding 状态：只接收 BPDU 报文。

端口状态和端口角色是没有必然联系的，各种端口角色能够具有的端口状态见表 10-1。

表 10-1 各种端口角色具有的端口状态

端口状态 \ 端口角色	根端口/Master 端口	指定端口	域边缘端口	Alternate 端口	Backup 端口
Forwarding	√	√	√	—	—
Learning	√	√	√	—	—
Discarding	√	√	√	√	√

3. 配置 MSTP 的工作模式

STP 兼容模式。交换机各个端口将发送 STP 报文。如果交换网络中存在运行 STP 协议的交换机，可以通过 stp mode stp 命令配置当前的 MSTP 运行在 STP 兼容模式下。

RSTP 兼容模式。交换机各个端口将发送 RSTP 报文。如果交换网络中存在运行 RSTP 协议的交换机，可以通过 stp mode rstp 命令配置当前的 MSTP 运行在 RSTP 兼容模式下。

MSTP 模式。交换机的各个端口将发送 MSTP 报文或者 STP 报文（如果端口上连接了 STP 交换机），并且具备多生成树的功能。

10.4 项目总结与提高

（1）写出主要项目实施规划、步骤与实训所得的主要结论。
（2）请课余时间详细研究 MSTP 的原理，设计实验验证该原理。

第二篇 路由部分

项目十一 OSPF 基本配置

11.1 项目提出

小王在某公司上班,发现在 RTA 上 ping PCB(IP 地址为 192.168.3.2),结果是无法互通,导致这种结果的原因是 RTA 上只有直连路由,没有到达 PCB 的路由表,故从 PCA 上来的数据报文无法转发给 PCB。他想用新学的 OSPF 协议试试,他该怎么办?

11.2 项目分析

1. 项目实训目的

掌握 H3C 路由器上配置 OSPF 的基本命令;

通过使用"debugging",能清晰地理解 OSPF 的几种协议报文:HELLO,DD,ACK,REQUEST 和 UPDATE 的主要功能;

理解 OSPF 邻居关系的建立过程以及路由器之间链路状态数据库(LSDB)的同步过程;

理解 OSPF 协议描述区域网络的方法;

掌握 OSPF 中基于端口参数的配置方法。

2. 项目实现功能

通过实训,理解 OSPF 的原理,并掌握其配置方法。

3. 项目主要应用的技术介绍

(1) OSPF。

OSPF(Open Shortest Path First,开放式最短路径优先)是一个内部网关协议(Interior Gateway Protocol,IGP),作用是在单一自治系统(Autonomous System,AS)内决策路由。它是对链路状态路由协议的一种实现,隶属内部网关协议(IGP),因此运作于自治系统内部。与 RIP 相比,OSPF 是链路状态协议,而 RIP 是距离矢量协议。

(2) Router-ID。

每一台 OSPF 路由器只有一个 Router-ID,Router-ID 使用 IP 地址的形式来表示,确定 Router-ID 的方法为:

- 手工指定 Router-ID。

- 路由器上活动 Loopback 接口中 IP 地址最大的,也就是数字最大的,如 C 类地址优先于 B 类地址,一个非活动的接口的 IP 地址是不能被选为 Router-ID 的。
- 如果没有活动的 Loopback 接口,则选择活动物理接口 IP 地址最大的。

(3) COST。

OSPF 使用接口的带宽来计算 Metric,OSPF 会自动计算接口上的 Cost 值,但也可以通过手工指定该接口的 Cost 值,手工指定优先于自动计算的值。

OSPF 计算的 Cost 和接口带宽成反比,带宽越高,Cost 值越小。到达目标相同 Cost 值的路径,可以执行负载均衡,最多有 6 条链路同时执行负载均衡。

(4) 链路(Link)。

链路就是路由器上的接口,在这里应该指运行在 OSPF 进程下的接口。

(5) 链路状态(Link-State)。

链路状态(LSA)就是 OSPF 接口上的描述信息,例如接口上的 IP 地址,子网掩码,网络类型,Cost 值等,OSPF 路由器之间交换的并不是路由表,而是链路状态(LSA),OSPF 通过获得网络中所有的链路状态信息,从而计算出到达每个目标精确的网络路径。OSPF 路由器会将自己的链路状态发给邻居,邻居将收到的链路状态全部放入链路状态数据库(Link-State Database),邻居再发给自己的所有邻居,并且在传递过程中绝对不会有任何更改。通过这样的过程,最终网络中所有的 OSPF 路由器都拥有网络中所有的链路状态,并且所有路由器的链路状态应该能描绘出相同的网络拓扑。

(6) OSPF 区域。

因为 OSPF 路由器之间会将所有的链路状态(LSA)相互交换,毫不保留,当网络规模达到一定程度时,LSA 将形成一个庞大的数据库,势必会给 OSPF 计算带来巨大的压力;为了能够降低 OSPF 计算的复杂程度,缓存计算压力,OSPF 采用分区域计算,将网络中所有 OSPF 路由器划分成不同的区域,每个区域负责各自区域精确的 LSA 传递与路由计算,然后再将一个区域的 LSA 简化和汇总之后转发到另外一个区域,这样,在区域内部,拥有网络精确的 LSA,而在不同区域,则传递简化的 LSA。

如果一台 OSPF 路由器属于单个区域,即该路由器所有接口都属于同一个区域,那么这台路由器称为 Internal Router(IR);如果一台 OSPF 路由器属于多个区域,即该路由器的接口不都属于一个区域,那么这台路由器称为 Area Border Router(ABR),ABR 可以将一个区域的 LSA 汇总后转发至另一个区域;如果一台 OSPF 路由器将外部路由协议重分布进 OSPF,那么这台路由器称为 Autonomous System Boundary Router(ASBR),如果只是将 OSPF 重分布进其他路由协议,则不能称为 ASBR。

(7) 邻居(Neighbor)。

OSPF 只有邻居之间才会交换 LSA,路由器会将链路状态数据库中所有的内容毫不保留地发给所有邻居,要想在 OSPF 路由器之间交换 LSA,必须先形成 OSPF 邻居,OSPF 邻居靠发送 Hello 包来建立和维护,Hello 包会在启动了 OSPF 的接口上周期性发送。

(8) Area-id(区域号码)。

路由器之间必须配置在相同的 OSPF 区域,否则无法形成邻居。

(9) DR/BDR。

当多台 OSPF 路由器连到同一个多路访问网段时,如果每两台路由器之间都相互交换

LSA，那么该网段将充满着众多 LSA 条目，为了能够尽量减少 LSA 的传播数量，通过在多路访问网段中选择出一个核心路由器，称为 DR（Designated Router），网段中所有的 OSPF 路由器都和 DR 互换 LSA，这样 DR 就会拥有所有的 LSA，并且将所有的 LSA 转发给每一台路由器；DR 就像是该网段的 LSA 中转站，所有的路由器都与该中转站互换 LSA，如果 DR 失效后，那么就会造成 LSA 的丢失与不完整，所以在多路访问网络中除了选举出 DR 之外，还会选举出一台路由器作为 DR 的备份，称为 BDR（Backup Designated Router）。BDR 在 DR 不可用时，代替 DR 的工作，而既不是 DR，也不是 BDR 的路由器称为 Drother，事实上，Dother 除了和 DR 互换 LSA 之外，同时还会和 BDR 互换 LSA。

11.3 项目实施

1．项目拓扑图

OSPF 基本配置如图 11-1 所示。

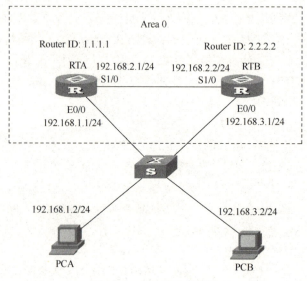

图 11-1　OSPF 基本配置

2．项目实训环境准备

MSR20-40（2 台），以太网交换机一台，计算机（2 台）、网线若干条，V.35 或 V.24DTE/DCE 线缆一对。CMW 版本：5.20。

为了不受原来的配置影响，在实训之前先将所有的配置数据擦除后重新启动，命令为：

    <h3c>Reset saved-configuration
    <h3c>reboot

3．项目主要实训步骤

（1）按图搭建实训环境并完成基本配置（包括 IP 地址配置）。
（2）检查网络连通性和路由器路由表。
在 RTA 上 ping PCB（IP 地址为 192.168.3.2），结果是无法互通，导致这种结果的原因是 RTA 上只有直连路由，没有到达 PCB 的路由表，故从 PCA 上来的数据报文无法转发给 PCB。

```
[RTA]display ip routing-table
Routing Tables: Public
         Destinations : 7        Routes : 7
Destination/Mask    Proto   Pre   Cost     NextHop         Interface
127.0.0.0/8         Direct  0     0        127.0.0.1       InLoop0
127.0.0.1/32        Direct  0     0        127.0.0.1       InLoop0
192.168.1.0/24      Direct  0     0        192.168.1.1     Eth0/0
192.168.1.1/32      Direct  0     0        127.0.0.1       InLoop0
192.168.2.0/24      Direct  0     0        192.168.2.1     S1/0
192.168.2.1/32      Direct  0     0        127.0.0.1       InLoop0
192.168.2.2/32      Direct  0     0        192.168.2.2     S1/0
```

（3）配置OSPF。

①在RTA上完成OSPF如下配置：

```
[RTA]ospf 1 router-id 192.168.1.1
```

如上配置中，数字1的含义是OSPF进程号，默认情况下取值为1。

```
[RTA-ospf-1]area 0.0.0.0
[RTA-ospf-1-area-0.0.0.0]network 192.168.1.0 0.0.0.255
[RTA-ospf-1-area-0.0.0.0]network 192.168.2.0 0.0.0.255
```

②在RTB上配置OSPF：

```
[RTB]ospf 1 router-id 192.168.2.2
[RTB-ospf-1]area 0.0.0.0
[RTB-ospf-1-area-0.0.0.0]network 192.168.2.0 0.0.0.255
[RTB-ospf-1-area-0.0.0.0]network 192.168.3.0 0.0.0.255
```

（4）在RTA上执行display OSPF brief，显示如下：

```
[RTA]dis ospf brief
         OSPF Process 1 with Router ID 192.168.1.1
                 OSPF Protocol Information

RouterID: 192.168.1.1        Border Router:
Route Tag: 0
Multi-VPN-Instance is not enabled
Applications Supported: MPLS Traffic-Engineering
SPF-schedule-interval: 5 0 5000
LSA generation interval: 5 0 5000
LSA arrival interval: 1000
Transmit pacing: Interval: 20 Count: 3
Default ASE parameters: Metric: 1 Tag: 1 Type: 2
Route Preference: 10
Route Preference: 10
SPF Computation Count: 5
RFC 1583 Compatible
Graceful restart interval: 120
Area Count: 1    Nssa Area Count: 0
ExChange/Loading Neighbors: 0
```

Area: 0.0.0.0   (MPLS TE not enabled)
Authtype: None Area flag: Normal
SPF Scheduled Count: 5
ExChange/Loading Neighbors: 0
Interface: 192.168.2.1 (Serial1/0) --> 192.168.2.2
Cost: 1562   State: P-2-P   Type: PTP   MTU: 1500
Timers: Hello 10, Dead 40, Poll 40, Retransmit 5, Transmit Delay 1
Interface: 192.168.1.1 (Ethernet0/0)
Cost: 1   State: DR   Type: Broadcast   MTU: 1500
Priority: 1
Designated Router: 192.168.1.1
Backup Designated Router: 0.0.0.0
Timers: Hello 10, Dead 40, Poll 40, Retransmit 5, Transmit Delay 1

（5）在 RTB 上执行 display OSPF brief，显示如下：

[RTB]dis ospf brief
        OSPF Process 1 with Router ID 192.168.2.2
            OSPF Protocol Information
RouterID: 192.168.2.2   Border Router:
Route Tag: 0
Multi-VPN-Instance is not enabled
Applications Supported: MPLS Traffic-Engineering
SPF-schedule-interval: 5 0 5000
LSA generation interval: 5 0 5000
LSA arrival interval: 1000
Transmit pacing: Interval: 20 Count: 3
Default ASE parameters: Metric: 1 Tag: 1 Type: 2
Route Preference: 10
ASE Route Preference: 150
SPF Computation Count: 4
RFC 1583 Compatible
Graceful restart interval: 120
Area Count: 1   Nssa Area Count: 0
ExChange/Loading Neighbors: 0
Area: 0.0.0.0   (MPLS TE not enabled)
Authtype: None Area flag: Normal
SPF Scheduled Count: 4
ExChange/Loading Neighbors: 0
Interface: 192.168.2.2 (Serial1/0) --> 192.168.2.1
Cost: 1562   State: P-2-P   Type: PTP   MTU: 1500
Timers: Hello 10, Dead 40, Poll 40, Retransmit 5, Transmit Delay 1
Interface: 192.168.3.1 (Ethernet0/0)
Cost: 1   State: DR   Type: Broadcast   MTU: 1500
Priority: 1
Designated Router: 192.168.3.1
Backup Designated Router: 0.0.0.0
Timers: Hello 10, Dead 40, Poll 40, Retransmit 5, Transmit Delay 1

注意：如果一台路由器没有手工配置 router id，那么它会如何选择 router id 呢？

如果一台路由器上没有手工配置 router ID，则系统会从当前接口的 IP 地址中自动选取一个。选取原则如下：如果路由器配置了 loopback 接口，则优选 loopback 接口；如果没有 loopback 接口，则从已经 up 的物理接口中选择接口 IP 地址最小的一个。

由于自动选举的 router id 会随着 IP 地址的变化而改变，这样干扰协议的正常运行，所以强烈建议：手工制定 router id。

（6）通过执行 display ospf lsdb 命令显示 OSPF 连接状态库的详细信息（即 RTA 和 RTB 发送的所有 LSA）。

```
[RTA]display ospf lsdb
          OSPF Process 1 with Router ID 192.168.1.1
                  Link State Database
                      Area: 0.0.0.0
 Type      LinkState ID    AdvRouter       Age   Len   Sequence    Metric
 Router    192.168.1.1     192.168.1.1     373   60    80000005    0
 Router    192.168.2.2     192.168.2.2     306   60    80000005    0
[RTB]dis ospf lsdb
          OSPF Process 1 with Router ID 192.168.2.2
                  Link State Database
                      Area: 0.0.0.0
 Type      LinkState ID    AdvRouter       Age   Len   Sequence    Metric
 Router    192.168.1.1     192.168.1.1     490   60    80000005    0
 Router    192.168.2.2     192.168.2.2     421   60    80000005    0
```

请对输出的信息给出确切的解释（这些信息对于路由器的日常管理工作非常重要）。请总结 OSPF 协议是怎样描述点对点网络和广播网络的。

下面是针对 OSPF 的五种协议报文的调试实训：

OSPF 的报文类型一共有五种报文类型：Hello 报文、DD 报文、REQUEST 报文、UPDATE 和 ACK 报文，下面可以用 Debugging OSPF Packet 命令打开所有五种报文的调试开关，或者用 Debugging OSPF Packet hello/DD/REQUEST/UPDATE/ACK 命令分别打开五种报文的调试开关，已进行观察。

（7）执行下面命令打开 HELLO 报文的调试信息：

```
<RTB>debugging ospf packet hello
<RTB>terminal monitor
<RTB>terminal debugging
```

仔细观察输出结果，然后试着解释一下本地路由器发送和接收的 HELLO 报文的每条参数的含义（完成本部分的实训后可以用 undo debugging all 命令来关闭调试开关）。

```
*Jul 30 14:50:09:44 2010 RTB RM/6/RMDEBUG:OSPF 1: SEND Packet.
*Jul 30 14:50:09:45 2010 RTB RM/6/RMDEBUG:Source Address: 192.168.3.1
*Jul 30 14:50:09:45 2010 RTB RM/6/RMDEBUG:Destination Address: 224.0.0.5
*Jul 30 14:50:09:45 2010 RTB RM/6/RMDEBUG:Ver# 2, Type: 1, Length: 44.
*Jul 30 14:50:09:45 2010 RTB RM/6/RMDEBUG:Router: 192.168.2.2, Area: 0.0.0.0, Checksum: 30282.
*Jul 30 14:50:09:46 2010 RTB RM/6/RMDEBUG:AuType: 00, Key(ascii): 0 0 0 0 0 0 0 0.
```

```
*Jul 30 14:50:09:46 2010 RTB RM/6/RMDEBUG:Net Mask: 255.255.255.0, Hello Int: 10, Option: _E_.
    *Jul 30 14:50:09:46 2010 RTB RM/6/RMDEBUG:Rtr Priority: 1, Dead Int: 40, DR: 192.168.3.1, BDR: 0.0.0.0.
    *Jul 30 14:50:13:44 2010 RTB RM/6/RMDEBUG:
    *Jul 30 14:50:13:44 2010 RTB RM/6/RMDEBUG:OSPF 1: SEND Packet.
    *Jul 30 14:50:13:45 2010 RTB RM/6/RMDEBUG:Source Address: 192.168.2.2
    *Jul 30 14:50:13:45 2010 RTB RM/6/RMDEBUG:Destination Address: 224.0.0.5
    *Jul 30 14:50:13:45 2010 RTB RM/6/RMDEBUG:Ver# 2, Type: 1, Length: 48.
    *Jul 30 14:50:13:46 2010 RTB RM/6/RMDEBUG:Router: 192.168.2.2, Area: 0.0.0.0, Checksum: 30790.
    *Jul 30 14:50:13:46 2010 RTB RM/6/RMDEBUG:AuType: 00, Key(ascii): 0 0 0 0 0 0 0 0.
    *Jul 30 14:50:13:46 2010 RTB RM/6/RMDEBUG:Net Mask: 255.255.255.0, Hello Int: 10, Option: _E_.
    *Jul 30 14:50:13:46 2010 RTB RM/6/RMDEBUG:Rtr Priority: 1, Dead Int: 40, DR: 0.0.0.0, BDR: 0.0.0.0.
    *Jul 30 14:50:13:197 2010 RTB RM/6/RMDEBUG:Attached Neighbor: 192.168.1.1.
    *Jul 30 14:50:14:670 2010 RTB RM/6/RMDEBUG:
    *Jul 30 14:50:14:671 2010 RTB RM/6/RMDEBUG:OSPF 1: RECV Packet.
    *Jul 30 14:50:14:671 2010 RTB RM/6/RMDEBUG:Source Address: 192.168.2.1
    *Jul 30 14:50:14:671 2010 RTB RM/6/RMDEBUG:Destination Address: 224.0.0.5
    *Jul 30 14:50:14:672 2010 RTB RM/6/RMDEBUG:Ver# 2, Type: 1, Length: 48.
    *Jul 30 14:50:14:672 2010 RTB RM/6/RMDEBUG:Router: 192.168.1.1, Area: 0.0.0.0, Checksum: 30790.
    *Jul 30 14:50:14:672 2010 RTB RM/6/RMDEBUG:AuType: 00, Key(ascii): 0 0 0 0 0 0 0 0.
    *Jul 30 14:50:14:673 2010 RTB RM/6/RMDEBUG:Net Mask: 255.255.255.0, Hello Int: 10, Option: _E_.
    *Jul 30 14:50:14:673 2010 RTB RM/6/RMDEBUG:Rtr Priority: 1, Dead Int: 40, DR: 0.0.0.0, BDR: 0.0.0.0.
    *Jul 30 14:50:14:823 2010 RTB RM/6/RMDEBUG:Attached Neighbor: 192.168.2.2.
```

RTA 和 RTB 在发现邻居之后交换 HELLO 报文，下一步双方开始发送各自的链路状态数据库。为了提高发送效率，双方需要先了解一下对端数据库中哪些 LSA 是自己需要的（如果某一条 LSA 自己已经有了，就不再需要了），方法是先发送 DD 报文。

请认真观察 DD 报文，并注意以下问题：

并不是所有的 DD 报文都具有相同的功能。比如终端上显示的前两个 DD 报文被用来在两个路由器之间协商某些参数，以建立一条可靠的连接。根据观察，请列出这些参数，并描述协商的过程。然后请说明后续的 DD 报文是如何利用这些参数的？

仔细分析报文中的 LSA 包头，并解释 LSA 包头在交换过程中的变化。

下面是对 DD 报文的分析，以供参考解答：

首先打开报文调试开关，并重启 OSPF 进程，通过终端捕获调试信息。

```
<RTB>debugging ospf packet dd
<RTB>terminal monitor
<RTB>terminal debugging
<RTB>reset ospf 1 process
```

输出信息如下：

Jul 30 15:23:01:992 2010 RTB RM/6/RMDEBUG:

*Jul 30 15:23:01:992 2010 RTB RM/6/RMDEBUG:OSPF 1: SEND Packet.
*Jul 30 15:23:01:993 2010 RTB RM/6/RMDEBUG:Source Address: 192.168.2.2
*Jul 30 15:23:01:993 2010 RTB RM/6/RMDEBUG:Destination Address: 224.0.0.5
*Jul 30 15:23:01:993 2010 RTB RM/6/RMDEBUG:Ver# 2, Type: 2, Length: 32.
*Jul 30 15:23:01:994 2010 RTB RM/6/RMDEBUG:Router: 192.168.2.2, Area: 0.0.0.0, Checksum: 13999.
*Jul 30 15:23:01:994 2010 RTB RM/6/RMDEBUG:AuType: 00, Key(ascii): 0 0 0 0 0 0 0 0.
*Jul 30 15:23:01:994 2010 RTB RM/6/RMDEBUG:MTU: 0, Option: _E_, R_I_M_MS Bit: _I_M_MS_.
*Jul 30 15:23:01:994 2010 RTB RM/6/RMDEBUG:DD SeqNumber: 27d.
*Jul 30 15:23:02:145 2010 RTB RM/6/RMDEBUG:

考虑到报文传输的可靠性，在 DD 报文的发送过程中需要确定双方的主从关系。上面的报文是 RTB 首先发送的一个 DD 报文，宣称自己是 Master（MS=1），并规定序列号为 27D。该报文不含 LSA 摘要信息。

%Jul 30 15:23:02:205 2010 RTB RM/3/RMLOG:OSPF-NBRCHANGE: Process 1, Neighbor 192.168.2.1(Serial1/0) from Loading to Full
*Jul 30 15:23:02:256 2010 RTB RM/6/RMDEBUG:OSPF 1: RECV Packet.
*Jul 30 15:23:02:356 2010 RTB RM/6/RMDEBUG:Source Address: 192.168.2.1
*Jul 30 15:23:02:407 2010 RTB RM/6/RMDEBUG:Destination Address: 224.0.0.5
*Jul 30 15:23:02:507 2010 RTB RM/6/RMDEBUG:Ver# 2, Type: 2, Length: 32.
*Jul 30 15:23:02:608 2010 RTB RM/6/RMDEBUG:Router: 192.168.1.1, Area: 0.0.0.0, Checksum: 14415.
*Jul 30 15:23:02:708 2010 RTB RM/6/RMDEBUG:AuType: 00, Key(ascii): 0 0 0 0 0 0 0 0.
*Jul 30 15:23:02:759 2010 RTB RM/6/RMDEBUG:MTU: 0, Option: _E_, R_I_M_MS Bit: _I_M_MS_.
*Jul 30 15:23:02:909 2010 RTB RM/6/RMDEBUG:DD SeqNumber: 1de.
*Jul 30 15:23:03:14 2010 RTB RM/6/RMDEBUG:

上面报文是 RTA 回应的一个 DD 报文（该报文同样不包含 LSA 的摘要信息），并规定 DD 序列号为 1DE。

*Jul 30 15:23:03:114 2010 RTB RM/6/RMDEBUG:OSPF 1: SEND Packet.
*Jul 30 15:23:03:175 2010 RTB RM/6/RMDEBUG:Source Address: 192.168.2.2
*Jul 30 15:23:03:225 2010 RTB RM/6/RMDEBUG:Destination Address: 224.0.0.5
*Jul 30 15:23:03:325 2010 RTB RM/6/RMDEBUG:Ver# 2, Type: 2, Length: 32.
*Jul 30 15:23:03:426 2010 RTB RM/6/RMDEBUG:Router: 192.168.2.2, Area: 0.0.0.0, Checksum: 13999.
*Jul 30 15:23:03:526 2010 RTB RM/6/RMDEBUG:AuType: 00, Key(ascii): 0 0 0 0 0 0 0 0.
*Jul 30 15:23:03:627 2010 RTB RM/6/RMDEBUG:MTU: 0, Option: _E_, R_I_M_MS Bit: _I_M_MS_.
*Jul 30 15:23:03:677 2010 RTB RM/6/RMDEBUG:DD SeqNumber: 27d.
*Jul 30 15:23:03:778 2010 RTB RM/6/RMDEBUG:

上面报文是 RTB 收到 RTA 的 DD 报文后，回应给 RTA 的报文，由于 RTB 的 router ID 是 192.168.2.2，比 RTA 的 router ID 192.168.1.1 要大，所以 RTB 报文认为自己是 Master，并重新规定序列号为 27D。

*Jul 30 15:23:03:828 2010 RTB RM/6/RMDEBUG:OSPF 1: RECV Packet.
*Jul 30 15:23:03:929 2010 RTB RM/6/RMDEBUG:Source Address: 192.168.2.1
*Jul 30 15:23:04:39 2010 RTB RM/6/RMDEBUG:Destination Address: 224.0.0.5
*Jul 30 15:23:04:90 2010 RTB RM/6/RMDEBUG:Ver# 2, Type: 2, Length: 72.

*Jul 30 15:23:04:190 2010 RTB RM/6/RMDEBUG:Router: 192.168.1.1, Area: 0.0.0.0, Checksum: 36289.

　　　　*Jul 30 15:23:04:291 2010 RTB RM/6/RMDEBUG:AuType: 00, Key(ascii): 0 0 0 0 0 0 0 0.
　　　　*Jul 30 15:23:04:341 2010 RTB RM/6/RMDEBUG:MTU: 0, Option: _E_, R_I_M_MS Bit: _M_.
　　　　*Jul 30 15:23:04:441 2010 RTB RM/6/RMDEBUG:DD SeqNumber: 27d.
　　　　*Jul 30 15:23:04:542 2010 RTB RM/6/RMDEBUG:LSAType: 1.
　　　　*Jul 30 15:23:04:592 2010 RTB RM/6/RMDEBUG:LinkStateId: 192.168.1.1.
　　　　*Jul 30 15:23:04:693 2010 RTB RM/6/RMDEBUG:Advertising Rtr: 192.168.1.1.
　　　　*Jul 30 15:23:04:743 2010 RTB RM/6/RMDEBUG:LSA Age: 5 Options: ExRouting:ON.
　　　　*Jul 30 15:23:04:854 2010 RTB RM/6/RMDEBUG:Length: 48 Seq# 80000007 CheckSum: 22296.
　　　　*Jul 30 15:23:04:954 2010 RTB RM/6/RMDEBUG:LSAType: 1.
　　　　*Jul 30 15:23:05:55 2010 RTB RM/6/RMDEBUG:LinkStateId: 192.168.2.2.
　　　　*Jul 30 15:23:05:105 2010 RTB RM/6/RMDEBUG:Advertising Rtr: 192.168.2.2.
　　　　*Jul 30 15:23:05:205 2010 RTB RM/6/RMDEBUG:LSA Age: 107 Options: ExRouting:ON.
　　　　*Jul 30 15:23:05:256 2010 RTB RM/6/RMDEBUG:Length: 60 Seq# 80000004 CheckSum: 17696.
　　　　*Jul 30 15:23:05:306 2010 RTB RM/6/RMDEBUG:

　　上面是 RTA 收到报文后，同意了 RTB 为 Master。RTA 使用了 RTB 的序列号 27D 发送的新的 DD 报文，且该报文开始正式传送 LSA 的摘要。

　　　　*Jul 30 15:23:05:407 2010 RTB RM/6/RMDEBUG:OSPF 1: SEND Packet.
　　　　*Jul 30 15:23:05:457 2010 RTB RM/6/RMDEBUG:Source Address: 192.168.2.2
　　　　*Jul 30 15:23:05:558 2010 RTB RM/6/RMDEBUG:Destination Address: 224.0.0.5
　　　　*Jul 30 15:23:05:618 2010 RTB RM/6/RMDEBUG:Ver# 2, Type: 2, Length: 52.
　　　　*Jul 30 15:23:05:719 2010 RTB RM/6/RMDEBUG:Router: 192.168.2.2, Area: 0.0.0.0, Checksum: 36306.
　　　　*Jul 30 15:23:05:769 2010 RTB RM/6/RMDEBUG:AuType: 00, Key(ascii): 0 0 0 0 0 0 0 0.
　　　　*Jul 30 15:23:05:869 2010 RTB RM/6/RMDEBUG:MTU: 0, Option: _E_, R_I_M_MS Bit: _MS_.
　　　　*Jul 30 15:23:05:970 2010 RTB RM/6/RMDEBUG:DD SeqNumber: 27e.
　　　　*Jul 30 15:23:06:70 2010 RTB RM/6/RMDEBUG:LSAType: 1.
　　　　*Jul 30 15:23:06:121 2010 RTB RM/6/RMDEBUG:LinkStateId: 192.168.2.2.
　　　　*Jul 30 15:23:06:221 2010 RTB RM/6/RMDEBUG:Advertising Rtr: 192.168.2.2.
　　　　*Jul 30 15:23:06:272 2010 RTB RM/6/RMDEBUG:LSA Age: 6 Options: ExRouting:ON.
　　　　*Jul 30 15:23:06:372 2010 RTB RM/6/RMDEBUG:Length: 36 Seq# 80000001 CheckSum: 41291.
　　　　*Jul 30 15:23:06:483 2010 RTB RM/6/RMDEBUG:

　　RTB 收到报文后，将 RTA 的邻居状态机改为 Exchange，并发送新的 DD 报文来描述自己的 LSA 摘要，需要注意的是：此时 RTB 已经将自己的序号改为 27D+1=27E。

　　（作为 Master 的一方定义了一个序列号 seq，每发送一个新的 DD 报文将 seq 加一。作为 Slave 的一方，每次收到上一个 Master 的 DD 报文中的 seq。实际这种序号机制是一种隐含的确认方法，请大家回忆一下 TCP 的序号确认机制）。

　　　　*Jul 30 15:23:07:699 2010 RTB RM/6/RMDEBUG:OSPF 1: RECV Packet.
　　　　*Jul 30 15:23:07:750 2010 RTB RM/6/RMDEBUG:Source Address: 192.168.2.1
　　　　*Jul 30 15:23:07:850 2010 RTB RM/6/RMDEBUG:Destination Address: 224.0.0.5
　　　　*Jul 30 15:23:07:901 2010 RTB RM/6/RMDEBUG:Ver# 2, Type: 2, Length: 72.
　　　　*Jul 30 15:23:07:951 2010 RTB RM/6/RMDEBUG:Router: 192.168.1.1, Area: 0.0.0.0, Checksum: 36289.

*Jul 30 15:23:08:62 2010 RTB RM/6/RMDEBUG:AuType: 00, Key(ascii): 0 0 0 0 0 0 0 0.
*Jul 30 15:23:08:112 2010 RTB RM/6/RMDEBUG:MTU: 0, Option: _E_, R_I_M_MS Bit: _M_.
*Jul 30 15:23:08:213 2010 RTB RM/6/RMDEBUG:DD SeqNumber: 27d.
*Jul 30 15:23:08:313 2010 RTB RM/6/RMDEBUG:LSAType: 1.
*Jul 30 15:23:08:413 2010 RTB RM/6/RMDEBUG:LinkStateId: 192.168.1.1.
*Jul 30 15:23:08:464 2010 RTB RM/6/RMDEBUG:Advertising Rtr: 192.168.1.1.
*Jul 30 15:23:08:564 2010 RTB RM/6/RMDEBUG:LSA Age: 5 Options: ExRouting:ON.
*Jul 30 15:23:08:615 2010 RTB RM/6/RMDEBUG:Length: 48 Seq# 80000007 CheckSum: 22296.
*Jul 30 15:23:08:715 2010 RTB RM/6/RMDEBUG:LSAType: 1.
*Jul 30 15:23:08:816 2010 RTB RM/6/RMDEBUG:LinkStateId: 192.168.2.2.
*Jul 30 15:23:08:927 2010 RTB RM/6/RMDEBUG:Advertising Rtr: 192.168.2.2.
*Jul 30 15:23:08:977 2010 RTB RM/6/RMDEBUG:LSA Age: 107 Options: ExRouting:ON.
*Jul 30 15:23:09:27 2010 RTB RM/6/RMDEBUG:Length: 60 Seq# 80000004 CheckSum: 17696.
*Jul 30 15:23:09:128 2010 RTB RM/6/RMDEBUG:

*Jul 30 15:23:09:178 2010 RTB RM/6/RMDEBUG:OSPF 1: RECV Packet.
*Jul 30 15:23:09:279 2010 RTB RM/6/RMDEBUG:Source Address: 192.168.2.1
*Jul 30 15:23:09:329 2010 RTB RM/6/RMDEBUG:Destination Address: 224.0.0.5
*Jul 30 15:23:09:430 2010 RTB RM/6/RMDEBUG:Ver# 2, Type: 2, Length: 32.
*Jul 30 15:23:09:481 2010 RTB RM/6/RMDEBUG:Router: 192.168.1.1, Area: 0.0.0.0, Checksum: 14262.
*Jul 30 15:23:09:581 2010 RTB RM/6/RMDEBUG:AuType: 00, Key(ascii): 0 0 0 0 0 0 0 0.
*Jul 30 15:23:09:642 2010 RTB RM/6/RMDEBUG:MTU: 0, Option: _E_, R_I_M_MS Bit: .
*Jul 30 15:23:09:742 2010 RTB RM/6/RMDEBUG:DD SeqNumber: 27e.
*Jul 30 15:23:09:843 2010 RTB RM/6/RMDEBUG:

*Jul 30 15:23:09:893 2010 RTB RM/6/RMDEBUG:OSPF 1: RECV Packet.
*Jul 30 15:23:09:994 2010 RTB RM/6/RMDEBUG:Source Address: 192.168.2.1
*Jul 30 15:23:10:94 2010 RTB RM/6/RMDEBUG:Destination Address: 224.0.0.5
*Jul 30 15:23:10:144 2010 RTB RM/6/RMDEBUG:Ver# 2, Type: 2, Length: 32.

11.4 项目总结与提高

（1）写出主要项目实施规划、步骤与实训所得的主要结论。
（2）请课余时间详细研究 OSPF 的原理。

项目十二 DR 的选举过程

12.1 项目提出

小王在某公司上班,想了解 OSPF 里 DR 的选举过程,于是他设计了如下实验。

12.2 项目分析

1. 项目实训目的

掌握 OSPF 中路由器优先级的设置;
理解选举 DR 的过程,以及怎样控制 DR 的选举。

2. 项目实现功能

通过实训,理解选举 DR 的过程,掌握 OSPF 中路由器优先级的设置。

3. 项目主要应用的技术介绍

DR/BDR 选举过程如下:

(1)在与一个或多个邻居之间的双向通信建立起来之后,路由器对每个邻居(发送来)的 Hello 包中的优先级、DR 和 BDR 域进行检查。列出所有能够参加选举的路由器(也就是说,路由器的优先级高于 0 并且此路由器的邻居状态至少为"双向");所有路由器都宣称自己为 DR(将它们自己的接口地址置于 Hello 包的 DR 域中);而且所有路由器都宣称自己为 BDR(将它们自己的接口地址置于 Hello 包的 BDR 域中)。进行计算的路由器也要将自身包括在此列表内,除非它被禁止参加选举。

(2)从以上备选路由器列表中,创造一个子集,此子集包含所有未宣称为 DR 的路由器(宣称自己为 DR 的路由器无法被选举为 BDR)。

(3)如果此子集中的一或多个邻居将它们自身的接口地址置于 BDR 域中,这些邻居中拥有最高优先级的路由器将被宣告为 BDR。如果出现平局(路由器优先级相等),拥有最高 Router ID 的邻居将被选举出来。

(4)如果此子集中没有任何路由器被宣告为 BDR,拥有最高优先级的邻居将被宣告为 BDR。如果出现平局,拥有最高 Router ID 的邻居将被选举出来。

(5)如果一或多个备选路由器将它们自身的接口地址置于 DR 域中,拥有最高优先级的邻居将被宣告为 DR。如果出现平局,拥有最高 Router ID 的邻居将被选举出来。

(6)如果没有任何路由器宣告自己为 DR,则新选举出来的 BDR 将成为 DR。

(7)如果进行计算的路由器是新选举出来的 DR 或者 BDR,或者它不再是 DR 或者 BDR,重复步骤(2)~(6)。

简而言之,当一个 OSPF 路由器启动并开始搜索邻居时,它先搜寻活动的 DR 和 BDR。如果 DR 和 BDR 存在,路由器就接受它们。如果没有 BDR,就进行一次选举将拥有最高优先级的路由器选举为 BDR。如果多于一台路由器拥有相同的优先级，那么拥有最高路由器 ID 的路由器将胜出。如果没有活动的 DR，BDR 将被提升为 DR 然后再进行一次 BDR 的选举。

12.3 项目实施

1．项目拓扑图

DR 的选举过程如图 12-1 所示。

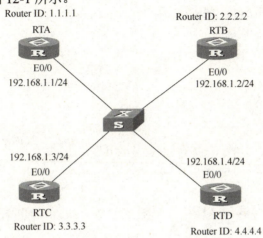

图 12-1　DR 的选举过程

2．项目实训环境准备

MSR20-40（4 台），以太网交换机一台，计算机（1 台）、网线若干条。CMW 版本：5.20。为了不受原来的配置影响，在实训之前先将所有的配置数据擦除后重新启动，命令为：

<h3c>Reset saved-configuration
<h3c>reboot

3．项目主要实训步骤

（1）按照实训拓扑图连接设备，给设备命名并设置 IP 地址（略）。为每台路由器手工设置 Router ID，并启动 OSPF 协议。

```
# 配置 RTA
<RTA> system-view
[RTA] router id 1.1.1.1
[RTA] ospf
[RTA-ospf-1] area 0
[RTA-ospf-1-area-0.0.0.0] network 192.168.1.0 0.0.0.255
[RTA-ospf-1-area-0.0.0.0] quit
[RTA-ospf-1] quit
# 配置 RTB。
<RTB> system-view
[RTB] router id 2.2.2.2
[RTB] ospf
[RTB-ospf-1] area 0
```

```
[RTB-ospf-1-area-0.0.0.0] network 192.168.1.0 0.0.0.255
[RTB-ospf-1-area-0.0.0.0] quit
[RTB-ospf-1] quit
# 配置 RouterC。
<RTC> system-view
[RTC] router id 3.3.3.3
[RTC] ospf
[RTC-ospf-1] area 0
[RTC-ospf-1-area-0.0.0.0] network 192.168.1.0 0.0.0.255
[RTC-ospf-1-area-0.0.0.0] quit
[RTC-ospf-1] quit
# 配置 RouterD。
<RTD> system-view
[RTD] router id 4.4.4.4
[RTD] ospf
[RTD-ospf-1]area 0
[RTD-ospf-1-area-0.0.0.0] network 192.168.1.0 0.0.0.255
[RTD-ospf-1-area-0.0.0.0] quit
[RTD-ospf-1]quit
```

(2) 重新启动 OSPF 进程。(想一想，为什么重新启动 OSPF 进程。)

```
<RTA>reset ospf 1 process
<RTB>reset ospf 1 process
<RTC>reset ospf 1 process
<RTD>reset ospf 1 process
```

(3) 在 RTA 上邻居的统计信息。

```
<RTA>display ospf peer
         OSPF Process 1 with Router ID 1.1.1.1
                Neighbor Brief Information
Area: 0.0.0.0
Router ID    Address        Pri    Dead-Time    Interface    State
2.2.2.2      192.168.1.2    1      33           Eth0/0       2-Way/ -
3.3.3.3      192.168.1.3    1      33           Eth0/0       Full/BDR
4.4.4.4      192.168.1.4    1      31           Eth0/0       Full/DR
```

可以看出，RTD 路由器是 DR，RTC 路由器是 BDR。

(4) 在 RTB 上邻居的统计信息。

```
<RTB>display ospf peer
         OSPF Process 1 with Router ID 2.2.2.2
                Neighbor Brief Information
Area: 0.0.0.0
Router ID    Address        Pri    Dead-Time    Interface    State
1.1.1.1      192.168.1.1    1      32           Eth0/0       2-Way/ -
3.3.3.3      192.168.1.3    1      33           Eth0/0       Full/BDR
4.4.4.4      192.168.1.4    1      31           Eth0/0       Full/DR
```

可以看出，与 RTA 显示的选举结果是一致的，这里路由器没有设置优先级，所以各路由器的优先级都是默认值 1。我们总结这样一原则：若路由器的 priority 值相等，则 Router ID 最大的当选为 DR，其次大的当选为 BDR。

（5）改变 RTA 和 RTB 的优先级如下：

[RTA-ethernet0/0]ospf dr-priority 3
[RTB-ethernet0/0] ospf dr-priority 2

（6）重新启动各路由器 OSPF 进程。

<RTA>reset ospf 1 process
<RTB>reset ospf 1 process
<RTC>reset ospf 1 process
<RTD>reset ospf 1 process
<RTD>dis ospf peer

OSPF Process 1 with Router ID 4.4.4.4
Neighbor Brief Information

Area: 0.0.0.0

Router ID	Address	Pri	Dead-Time	Interface	State
1.1.1.1	192.168.1.1	3	35	Eth0/0	Full/DR
2.2.2.2	192.168.1.2	2	34	Eth0/0	Full/BDR
3.3.3.3	192.168.1.3	1	32	Eth0/0	2-Way/ -

从上面显示可以看出，DR 和 BDR 都已经改变，RTA（Router ID 为 1.1.1.1）为 DR，RTB（Router ID 为 2.2.2.2）为 BDR。

（7）把 RTA 的 E0/0 接口 shutdown 掉，从 RTC 观察邻居关系的变化。

<RTC>dis ospf peer

OSPF Process 1 with Router ID 3.3.3.3
Neighbor Brief Information

Area: 0.0.0.0

Router ID	Address	Pri	Dead-Time	Interface	State
2.2.2.2	192.168.1.2	2	32	Eth0/0	Full/DR
4.4.4.4	192.168.1.4	1	30	Eth0/0	Full/BDR

先从上面显示可以看出：RTA 被 shutdown 掉之后，作为 DBR 的 RTB 马上接替了"DR"的位置。这就是 DR 快速响应的原则，BDR 的作用也体现在这里。

现在 undo shutdown RTA 的 E0/0 接口，RTA 又回到网络中，由于 RTA 的优先级比现任的 DR 要高，想一想：RTA 会不会抢回 DR 呢？

看一看显示结果：

<RTC>dis ospf peer

OSPF Process 1 with Router ID 3.3.3.3
Neighbor Brief Information

Area: 0.0.0.0

Router ID	Address	Pri	Dead-Time	Interface	State
1.1.1.1	192.168.1.1	3	38	Eth0/0	2-Way/ -
2.2.2.2	192.168.1.2	2	31	Eth0/0	Full/DR
4.4.4.4	192.168.1.4	1	39	Eth0/0	Full/BDR

事实证明：答案是否定的。由于 OSPF 遵循"稳定压倒一切的原则"，每一台新加入的路由器并不急于参加选举，而是先考察本网段中是否有 DR 存在。如果目前网段中已经存在 DR，即使本路由器的 priority 比现有的 DR 还高，也不会声称自己是 DR。

现在大家知道为什么观察 DR 选举过程是必须重新启动 OSPF 进程或重新启动路由器了吧！

12.4　项目总结与提高

写出主要项目实施规划、步骤与实训所得的主要结论。

项目十三　虚链路、路由聚合和路由引入

13.1　项目提出

小王在某公司上班,想了解虚链路、路由聚合和路由引入,于是他设计了如下实验。

13.2　项目分析

1. 项目实训目的

掌握 OSPF 虚链路的配置;
掌握有关路由聚合和路由引入的简单配置。

2. 项目实现功能

通过实训,掌握 OSPF 虚链路的配置,掌握有关路由聚合和路由引入的简单配置。

3. 项目主要应用的技术介绍

(1) 虚连接或者虚链路 (Virtual-link)。

虚连接是设置在两个路由器之间,这两个路由器都有一个端口与同一个非主干区域相连。虚连接被认为是属于主干区域的,在 OSPF 路由协议看来,虚连接两端的两个路由器被一个点对点的链路连接在一起。在 OSPF 路由协议中,通过虚连接的路由信息是作为域内路由来看待的。

虚连接是在两台 ABR 之间,穿过一个非骨干区域(转换区域——Transit Area),建立的一条逻辑上的连接通道,可以理解为两台 ABR 之间存在一个点对点的连接。"逻辑通道"是指两台 ABR 之间的多台运行 OSPF 的路由器只是起到一个转发报文的作用(由于协议报文的目的地址不是这些路由器,所以这些报文对于它们是透明的,只是当作普通的 IP 报文来转发),两台 ABR 之间直接传递路由信息。这里的路由信息是指由 ABR 生成的 type3 的 LSA,区域内的路由器同步方式没有因此改变。

(2) 路由聚合。

将多条路由合并成一条路由,通常在 ABR 上实现。虽然路由聚合可以在任意两个区域之间进行,但推荐在往骨干区的方向上进行。这样,骨干区会接收到所有聚合的路由,然后依次将聚合过的路由引入其他区域。

(3) 路由引入。

路由引入是说可以通过在某一个路由协议的模式下,通过引入其他协议的路由信息的方式来获得对方的路由信息。

13.3 项目实施

1. 项目拓扑图

虚链路、路由聚合和路由引入如图 13-1 所示。

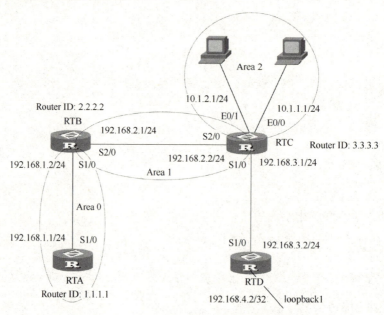

图 13-1 虚链路、路由聚合和路由引入

2. 项目实训环境准备

MSR20-40（4台），计算机（2台）、网线若干条。CMW 版本：5.20。

为了不受原来的配置影响，在实训之前先将所有的配置数据擦除后重新启动，命令为：

<h3c>Reset saved-configuration
<h3c>reboot

3. 项目主要实训步骤

任务一：虚链路的配置

（1）按照实训拓扑图连接设备，给设备命名并设置 IP 地址。

①RTA 的配置。

```
<H3C>sys
System View: return to User View with Ctrl+Z.
[H3C]sysn RTA
[RTA]dis ip int b
*down: administratively down
(s): spoofing
Interface                    Physical   Protocol   IP Address
Aux0                         down       down       unassigned
Ethernet0/0                  down       down       unassigned
Ethernet0/1                  down       down       unassigned
```

Serial1/0	up	up	unassigned
Serial2/0	down	down	unassigned

[RTA]int s1/0
[RTA-Serial1/0]ip add 192.168.1.1 24
[RTA-Serial1/0]

②RTB 的配置。

<H3C>sys
System View: return to User View with Ctrl+Z.
[H3C]sysn RTB
[RTB]dis ip int b
*down: administratively down
(s): spoofing

Interface	Physical	Protocol	IP Address	Description
Aux0	down	down	unassigned	Aux0 Inte...
Ethernet0/0	down	down	unassigned	Ethernet0...
Ethernet0/1	down	down	unassigned	Ethernet0...
Serial1/0	up	up	unassigned	Serial1/0...
Serial2/0	up	up	unassigned	Serial2/0...

[RTB]int s1/0
[RTB-Serial1/0]ip add 192.168.1.2 24
[RTB-Serial1/0]
%Aug 3 14:06:38:266 2010 RTB IFNET/4/UPDOWN:
    Protocol PPP IPCP on the interface Serial1/0 is UP
[RTB-Serial1/0]int s2/0
[RTB-Serial2/0]ip add 192.168.2.1 24
[RTB-Serial2/0]

③RTC 的配置。

<H3C>sys
System View: return to User View with Ctrl+Z.
[H3C]sysn RTC
[RTC]dis ip int b
*down: administratively down
(s): spoofing

Interface	Physical	Protocol	IP Address	Description
Aux0	down	down	unassigned	Aux0 Inte...
Ethernet0/0	up	down	unassigned	Ethernet0...
Ethernet0/1	up	down	unassigned	Ethernet0...
Serial1/0	up	up	unassigned	Serial1/0...
Serial2/0	up	up	unassigned	Serial2/0...

可检查线缆是否连接好。

[RTC]
[RTC]int e0/0
[RTC-Ethernet0/0]ip add 10.1.1.1 24
[RTC-Ethernet0/0]

```
%Aug   3 13:58:41:325 2010 RTC IFNET/4/UPDOWN:
   Line protocol on the interface Ethernet0/0 is UP
[RTC-Ethernet0/0]int e0/1
[RTC-Ethernet0/1]ip add 10.1.2.1 24
[RTC-Ethernet0/1]
%Aug   3 13:59:02:109 2010 RTC IFNET/4/UPDOWN:
   Line protocol on the interface Ethernet0/1 is UP
[RTC-Ethernet0/1]int s2/0
[RTC-Serial2/0]ip add 192.168.2.2 24
[RTC-Serial2/0]
%Aug   3 14:00:54:893 2010 RTC IFNET/4/UPDOWN:
   Protocol PPP IPCP on the interface Serial2/0 is UP
[RTC-Serial2/0]dis ip int b
*down: administratively down
(s): spoofing
```

Interface	Physical	Protocol	IP Address	Description
Aux0	down	down	unassigned	Aux0 Inte...
Ethernet0/0	up	up	10.1.1.1	Ethernet0...
Ethernet0/1	up	up	10.1.2.1	Ethernet0...
Serial1/0	up	up	unassigned	Serial1/0...
Serial2/0	up	up	192.168.2.2	Serial2/0...

```
[RTC-Serial2/0]
```

在虚链路实训中，不会用到 RTD，所以现在不用配 Serial 1/0 接口地址。

（2）在各接口上配置 Router ID，运行 OSPF 协议，并划分成三个区域——Area 0、Area 1 和 Area2。

参考配置如下：

①RTA 的配置。

```
[RTA]router id 1.1.1.1
%Aug   3 14:03:58:914 2010 RTA RM/3/RMLOG:OSPF 1 226: New Router ID elected, plea
se restart OSPF if you want to make the new Router ID take effect.
[RTA]ospf 1
[RTA-ospf-1]area 0
[RTA-ospf-1-area-0.0.0.0]network 192.168.1.0 0.0.0.255
[RTA-ospf-1-area-0.0.0.0]
```

②RTB 的配置。

```
[RTB]router id 2.2.2.2
[RTB]ospf 1
[RTB-ospf-1]area 0
[RTB-ospf-1-area-0.0.0.0]network 192.168.1.0 0.0.0.255
[RTB-ospf-1-area-0.0.0.0]quit
[RTB-ospf-1]area 1
[RTB-ospf-1-area-0.0.0.1]network 192.168.2.0 0.0.0.255
[RTB-ospf-1-area-0.0.0.1]
```

③RTC 的配置。

[RTC]router id 3.3.3.3
[RTC]ospf 1
[RTC-ospf-1]area 1
[RTC-ospf-1-area-0.0.0.1]network 192.168.2.0 0.0.0.255
[RTC-ospf-1-area-0.0.0.1]quit
[RTC-ospf-1]area 2
[RTC-ospf-1-area-0.0.0.2]network 10.1.1.0 0.0.0.255
[RTC-ospf-1-area-0.0.0.2]network 10.1.2.0 0.0.0.255
[RTC-ospf-1-area-0.0.0.2]

(3) 在各路由器上用 display ip routing-table 命令查看路由表，信息如下：

[RTA]display ip routing-table
Routing Tables: Public
            Destinations : 6   Routes : 6
Destination/Mask   Proto   Pre   Cost   NextHop   Interface
127.0.0.0/8   Direct 0   0   127.0.0.1   InLoop0
127.0.0.1/32   Direct 0   0   127.0.0.1   InLoop0
192.168.1.0/24   Direct 0   0   192.168.1.1   S1/0
192.168.1.1/32   Direct 0   0   127.0.0.1   InLoop0
192.168.1.2/32   Direct 0   0   192.168.1.2   S1/0
192.168.2.0/24   OSPF   10   3124   192.168.1.2   S1/0

[RTB]display ip routing-table
Routing Tables: Public
            Destinations : 8   Routes : 8
Destination/Mask   Proto   Pre   Cost   NextHop   Interface
127.0.0.0/8   Direct 0   0   127.0.0.1   InLoop0
127.0.0.1/32   Direct 0   0   127.0.0.1   InLoop0
192.168.1.0/24   Direct 0   0   192.168.1.2   S1/0
192.168.1.1/32   Direct 0   0   192.168.1.1   S1/0
192.168.1.2/32   Direct 0   0   127.0.0.1   InLoop0
192.168.2.0/24   Direct 0   0   192.168.2.1   S2/0
192.168.2.1/32   Direct 0   0   127.0.0.1   InLoop0
192.168.2.2/32   Direct 0   0   192.168.2.2   S2/0

[RTC] display ip routing-table
Routing Tables: Public
            Destinations : 10   Routes : 10
Destination/Mask   Proto   Pre   Cost   NextHop   Interface
10.1.1.0/24   Direct 0   0   10.1.1.1   Eth0/0
10.1.1.1/32   Direct 0   0   127.0.0.1   InLoop0
10.1.2.0/24   Direct 0   0   10.1.2.1   Eth0/1
10.1.2.1/32   Direct 0   0   127.0.0.1   InLoop0
127.0.0.0/8   Direct 0   0   127.0.0.1   InLoop0
127.0.0.1/32   Direct 0   0   127.0.0.1   InLoop0
192.168.1.0/24   OSPF   10   3124   192.168.2.1   S2/0
192.168.2.0/24   Direct 0   0   192.168.2.2   S2/0
192.168.2.1/32   Direct 0   0   192.168.2.1   S2/0

192.168.2.2/32	Direct	0	0	127.0.0.1	InLoop0

从上面路由表中可以看出，只有从 Area 0 到 Area 1 和 Area1 到 Area 0 的 OSPF 路由，而没有 Area 0 到 Area 2 的路由，Area 2 的路由也不会传到 Area 1 和 Area0。

在 RTC 上直接执行 ping 192.168.1.1，信息如下：

```
[RTC]ping 192.168.1.1
  PING 192.168.1.1: 56    data bytes, press CTRL_C to break
    Reply from 192.168.1.1: bytes=56 Sequence=1 ttl=254 time=52 ms
    Reply from 192.168.1.1: bytes=56 Sequence=2 ttl=254 time=52 ms
    Reply from 192.168.1.1: bytes=56 Sequence=3 ttl=254 time=52 ms
    Reply from 192.168.1.1: bytes=56 Sequence=4 ttl=254 time=51 ms
    Reply from 192.168.1.1: bytes=56 Sequence=5 ttl=254 time=52 ms
  --- 192.168.1.1 ping statistics ---
    5 packet(s) transmitted
    5 packet(s) received
    0.00% packet loss
    round-trip min/avg/max = 51/51/52 ms
```

可以 ping 通，因为直接 ping 的话是以 192.168.2.2 为源地址的，然后在 RTC 上执行 ping –a 10.1.1.1 192.168.1.1 命令，看看结果是否能 ping 通。

```
[RTC]ping -a 10.1.1.1 192.168.1.1
  PING 192.168.1.1: 56    data bytes, press CTRL_C to break
    Request time out
    Request time out
    Request time out
    Request time out
```

请大家思考一下，为什么会出现这种情况呢？

这是因为在 OSPF 中，需要通过骨干区域来转发区域间的路由信息。所以就有这样一个要求：所有区域必须和骨干区域相连，也就是，每一个 ABR 连接区域中至少有一个骨干区域。但由于网络的拓扑结构复杂，有时无法满足每个区域必须和骨干区域直接相连的要求。本实训中的 Area 2 就是这种情况。为解决此问题，OSPF 提出了虚连接的概念，在本实训中需要建立一条虚链路来提供 Area 2 和 Area 0 的连接。

（4）虚链路的配置如下：

```
[RTB-ospf-1-area-0.0.0.1]vlink-peer 3.3.3.3
[RTC-ospf-1-area-0.0.0.1]vlink-peer 2.2.2.2
```

（5）重新启动接口，查看 RTA 和 RTB 的路由表。

```
[RTA]dis ip rou
Routing Tables: Public
         Destinations : 8        Routes : 8
Destination/Mask    Proto   Pre  Cost         NextHop          Interface
10.1.1.0/24         OSPF    10   3125         192.168.1.2      S1/0
10.1.2.0/24         OSPF    10   3125         192.168.1.2      S1/0
127.0.0.0/8         Direct  0    0            127.0.0.1        InLoop0
```

127.0.0.1/32	Direct	0	0	127.0.0.1	InLoop0
192.168.1.0/24	Direct	0	0	192.168.1.1	S1/0
192.168.1.1/32	Direct	0	0	127.0.0.1	InLoop0
192.168.1.2/32	Direct	0	0	192.168.1.2	S1/0
192.168.2.0/24	OSPF	10	3124	192.168.1.2	S1/0

[RTB]dis ip rou
Routing Tables: Public
        Destinations : 10   Routes : 10

Destination/Mask	Proto	Pre	Cost	NextHop	Interface
10.1.1.0/24	OSPF	10	1563	192.168.2.2	S2/0
10.1.2.0/24	OSPF	10	1563	192.168.2.2	S2/0
127.0.0.0/8	Direct	0	0	127.0.0.1	InLoop0
127.0.0.1/32	Direct	0	0	127.0.0.1	InLoop0
192.168.1.0/24	Direct	0	0	192.168.1.2	S1/0
192.168.1.1/32	Direct	0	0	192.168.1.1	S1/0
192.168.1.2/32	Direct	0	0	127.0.0.1	InLoop0
192.168.2.0/24	Direct	0	0	192.168.2.1	S2/0
192.168.2.1/32	Direct	0	0	127.0.0.1	InLoop0
192.168.2.2/32	Direct	0	0	192.168.2.2	S2/0

与配置虚链路之前的路由表对比一下，现在是不是已经有了 Area 2 的路由了，在 RTC 上执行"ping –a 10.1.1.1 192.168.1.1"命令。

```
[RTC]ping -a 10.1.1.1 192.168.1.1
    PING 192.168.1.1: 56    data bytes, press CTRL_C to break
      Reply from 192.168.1.1: bytes=56 Sequence=1 ttl=254 time=52 ms
      Reply from 192.168.1.1: bytes=56 Sequence=2 ttl=254 time=51 ms
      Reply from 192.168.1.1: bytes=56 Sequence=3 ttl=254 time=52 ms
      Reply from 192.168.1.1: bytes=56 Sequence=4 ttl=254 time=52 ms
      Reply from 192.168.1.1: bytes=56 Sequence=5 ttl=254 time=51 ms
    --- 192.168.1.1 ping statistics ---
      5 packet(s) transmitted
      5 packet(s) received
      0.00% packet loss
      round-trip min/avg/max = 51/51/52 ms
```

信息显示能够 ping 通，虚链路已经起作用了。

任务二：配置路由聚合

（1）本实训中区域 2 中有两条区域内路由 10.1.1.0/24 和 10.1.2.0/24，在 RTB 上会生成两条 LSA，我们可以通过路由聚合将两条路由聚合成 10.1.0.0/16。（注意：路由聚合只有在 ABR 上配置才会有效）我们只需在 RTC 上配置如下命令：

```
[RTC]ospf
[RTC-ospf-1]area 2
[RTC-ospf-1-area-0.0.0.2]abr-summary 10.1.0.0 255.255.0.0
```

（2）在 RTA 和 RTB 上查看聚合后的路由表如下：

[RTA]display ip routing-table

Routing Tables: Public
         Destinations : 7   Routes : 7

Destination/Mask	Proto	Pre	Cost	NextHop	Interface
10.1.0.0/16	OSPF	10	3125	192.168.1.2	S1/0
127.0.0.0/8	Direct	0	0	127.0.0.1	InLoop0
127.0.0.1/32	Direct	0	0	127.0.0.1	InLoop0
192.168.1.0/24	Direct	0	0	192.168.1.1	S1/0
192.168.1.1/32	Direct	0	0	127.0.0.1	InLoop0
192.168.1.2/32	Direct	0	0	192.168.1.2	S1/0
192.168.2.0/24	OSPF	10	3124	192.168.1.2	S1/0

[RTB] display ip routing-table
Routing Tables: Public
         Destinations : 9   Routes : 9

Destination/Mask	Proto	Pre	Cost	NextHop	Interface
10.1.0.0/16	OSPF	10	1563	192.168.2.2	S2/0
127.0.0.0/8	Direct	0	0	127.0.0.1	InLoop0
127.0.0.1/32	Direct	0	0	127.0.0.1	InLoop0
192.168.1.0/24	Direct	0	0	192.168.1.2	S1/0
192.168.1.1/32	Direct	0	0	192.168.1.1	S1/0
192.168.1.2/32	Direct	0	0	127.0.0.1	InLoop0
192.168.2.0/24	Direct	0	0	192.168.2.1	S2/0
192.168.2.1/32	Direct	0	0	127.0.0.1	InLoop0
192.168.2.2/32	Direct	0	0	192.168.2.2	S2/0

任务三：路由引入

按照实训拓扑图配置 RTC 和 RTD 各接口 IP 地址。

①配置 RTC 接口地址。

```
[RTC]dis ip in b
*down: administratively down
(s): spoofing
Interface              Physical    Protocol IP Address    Description
Aux0                   down        down     unassigned    Aux0 Inte...
Ethernet0/0            up          up       10.1.1.1      Ethernet0...
Ethernet0/1            up          up       10.1.2.1      Ethernet0...
Serial1/0              up          up       unassigned    Serial1/0...
Serial2/0              up          up       192.168.2.2   Serial2/0...
[RTC]int s1/0
[RTC-Serial1/0]ip add 192.168.3.1 24
[RTC-Serial1/0]
```

②配置 RTD 接口 IP 和默认路由。

```
<H3C>
#Aug  3 15:16:23:951 2010 H3C SHELL/4/LOGIN:
 Trap 1.3.6.1.4.1.25506.2.2.1.1.3.0.1<hh3cLogIn>: login from Console
%Aug  3 15:16:23:951 2010 H3C SHELL/4/LOGIN: Console login from con0
<H3C>sys
System View: return to User View with Ctrl+Z.
```

```
[H3C]sysn RTD
[RTD]dis ip int b
*down: administratively down
(s): spoofing
Interface                Physical  Protocol  IP Address    Description
Aux0                     down      down      unassigned    Aux0 Inte...
Ethernet0/0              down      down      unassigned    Ethernet0...
Ethernet0/1              down      down      unassigned    Ethernet0...
Serial1/0                up        up        unassigned    Serial1/0...
Serial2/0                down      down      unassigned    Serial2/0...
[RTD]int s1/0
[RTD-Serial1/0]ip add 192.168.3.2 24
[RTD-Serial1/0]
[RTD-Serial1/0]ip add 192.168.3.2 24
%Aug  3 15:19:58:483 2010 RTD IFNET/4/UPDOWN:
  Protocol PPP IPCP on the interface Serial1/0 is UP
[RTD-Serial1/0]int loopback1
[RTD-LoopBack1]ip add 192.168.4.2 32
[RTD-LoopBack1]quit
[RTD]ip route-static 0.0.0.0 0.0.0.0 192.168.3.1 preference 60
[RTC]ip route-static 192.168.4.0 24 192.168.3.2 preference 60
[RTC-ospf-1]import-route static   type  2 cost 10 tag 1
[RTA]dis ip rou
Routing Tables: Public
         Destinations : 8       Routes : 8
Destination/Mask    Proto  Pre   Cost      NextHop        Interface
10.1.0.0/16         OSPF   10    3125      192.168.1.2    S1/0
127.0.0.0/8         Direct 0     0         127.0.0.1      InLoop0
127.0.0.1/32        Direct 0     0         127.0.0.1      InLoop0
192.168.1.0/24      Direct 0     0         192.168.1.1    S1/0
192.168.1.1/32      Direct 0     0         127.0.0.1      InLoop0
192.168.1.2/32      Direct 0     0         192.168.1.2    S1/0
192.168.2.0/24      OSPF   10    3124      192.168.1.2    S1/0
192.168.4.0/24      O_ASE  150   10        192.168.1.2    S1/0
[RTB]dis ip rou
Routing Tables: Public
         Destinations : 10      Routes : 10
Destination/Mask    Proto  Pre   Cost      NextHop        Interface
10.1.0.0/16         OSPF   10    1563      192.168.2.2    S2/0
127.0.0.0/8         Direct 0     0         127.0.0.1      InLoop0
127.0.0.1/32        Direct 0     0         127.0.0.1      InLoop0
192.168.1.0/24      Direct 0     0         192.168.1.2    S1/0
192.168.1.1/32      Direct 0     0         192.168.1.1    S1/0
192.168.1.2/32      Direct 0     0         127.0.0.1      InLoop0
192.168.2.0/24      Direct 0     0         192.168.2.1    S2/0
192.168.2.1/32      Direct 0     0         127.0.0.1      InLoop0
192.168.2.2/32      Direct 0     0         192.168.2.2    S2/0
192.168.4.0/24      O_ASE  150   10        192.168.2.2    S2/0
```

现在我们发现路由表中已经引入了到 192.168.4.0 网段的路由，在从 RTA 上 ping192.168.4.2，通信正常，显示信息如下：

    [RTA]ping 192.168.4.2
      PING 192.168.4.2: 56   data bytes, press CTRL_C to break
        Reply from 192.168.4.2: bytes=56 Sequence=1 ttl=253 time=77 ms
        Reply from 192.168.4.2: bytes=56 Sequence=2 ttl=253 time=77 ms
        Reply from 192.168.4.2: bytes=56 Sequence=3 ttl=253 time=77 ms
        Reply from 192.168.4.2: bytes=56 Sequence=4 ttl=253 time=77 ms
        Reply from 192.168.4.2: bytes=56 Sequence=5 ttl=253 time=77 ms
      --- 192.168.4.2 ping statistics ---
        5 packet(s) transmitted
        5 packet(s) received
        0.00% packet loss
    round-trip min/avg/max = 77/77/77 ms

我们回到 RTD，执行命令 ping 192.168.1.1，显示信息如下：

    [RTD]ping 192.168.1.1
      PING 192.168.1.1: 56   data bytes, press CTRL_C to break
        Request time out
        Request time out
        Request time out
        Request time out
        Request time out
      --- 192.168.1.1 ping statistics ---
        5 packet(s) transmitted
        0 packet(s) received
        100.00% packet loss

又 ping 不通了，想想为什么？我们试一下命令 "ping –a 192.168.4.2 192.168.1.1"，显示信息如下：

    [RTD]ping -a 192.168.4.2 192.168.1.1
      PING 192.168.1.1: 56   data bytes, press CTRL_C to break
        Reply from 192.168.1.1: bytes=56 Sequence=1 ttl=253 time=77 ms
        Reply from 192.168.1.1: bytes=56 Sequence=2 ttl=253 time=76 ms
        Reply from 192.168.1.1: bytes=56 Sequence=3 ttl=253 time=77 ms
        Reply from 192.168.1.1: bytes=56 Sequence=4 ttl=253 time=78 ms
        Reply from 192.168.1.1: bytes=56 Sequence=5 ttl=253 time=76 ms
      --- 192.168.1.1 ping statistics ---
        5 packet(s) transmitted
        5 packet(s) received
        0.00% packet loss
      round-trip min/avg/max = 76/76/78 ms

为什么会这样呢？

这是因为在 RTD 上直接 ping 192.168.1.1，实际上是 s1/0（192.168.3.2）发出的，我们并没有引进该网段的路由，所以不会通。同理，在 RTA 上也 ping 不通 192.168.3.2。

对于这种情况，我们同样要引入 192.168.3.0/24 网段的路由，问题就可以解决。

    [RTC-ospf-1]import-route   direct
    [RTD]ping 192.168.1.1
      PING 192.168.1.1: 56   data bytes, press CTRL_C to break
        Reply from 192.168.1.1: bytes=56 Sequence=1 ttl=253 time=77 ms
        Reply from 192.168.1.1: bytes=56 Sequence=2 ttl=253 time=77 ms
        Reply from 192.168.1.1: bytes=56 Sequence=3 ttl=253 time=77 ms
        Reply from 192.168.1.1: bytes=56 Sequence=4 ttl=253 time=77 ms
        Reply from 192.168.1.1: bytes=56 Sequence=5 ttl=253 time=77 ms
      --- 192.168.1.1 ping statistics ---
      5 packet(s) transmitted
      5 packet(s) received
      0.00% packet loss
      round-trip min/avg/max = 77/77/77 m

引入直连，问题得以解决。

13.4　项目总结与提高

写出主要项目实施规划、步骤与实训所得的主要结论。

项目十四 OSPF 的默认路由

14.1 项目提出

小王在某公司上班,想了解 OSPF 的默认路由,于是他设计了如下实验。

14.2 项目分析

1. 项目实训目的

理解 OSPF 协议默认路由的产生过程。

2. 项目实现功能

通过实训,理解 OSPF 协议默认路由的产生过程。

3. 项目主要应用的技术介绍

OSPF 在不同类型的区域中引入默认路由,OSPF 默认路由产生和通告的方式是不同的,所以在介绍 OSPF 默认路由之前,先从 OSPF 的区域类型展开介绍。

(1) OSPF 区域类型。

OSPF 根据网络的需求可以定义为下列几种类型:普通区域,STUB 区域,完全 STUB 区域,NSSA 区域,完全 NSSA 区域。

①普通区域。

当区域被默认定义时,它被认为是普通区域。普通区域可以是标准区域或骨干区域。标准区域是最通用的区域,它携带区域内路由,区域间路由和外部路由。骨干区域是连接所有其他 OSPF 区域的中央区域。

②STUB 区域。

STUB 区域是一个不允许 AS 外部 LSA 在其内部泛洪的区域。STUB 区域只可以携带区域内路由和区域间路由。在这些区域中路由器的 OSPF 数据库和路由表规模以及路由信息传递的数量都会大大减少,为了保证到自治系统外的路由依旧可达,由该区域的 ABR 生成一条默认路由 0.0.0.0 传播到区域内,所有到自治系统外部的路由都必须通过 ABR 才能到达。

③完全 STUB 区域。

完全 STUB 区域是区域中最受限的形式,它不仅不允许携带外部路由,甚至连区域间路由也不允许携带,只可以携带区域内路由。在这些区域中路由器的 OSPF 数据库和路由表规模以及路由信息传递的数量都会大大减少,为了保证到区域外的路由依旧可达,由该区域的 ABR 生成一条默认路由 0.0.0.0 传播到区域内,所有到该区域外部的路由都必须通过 ABR 才能到达。

④NSSA 区域。

NSSA 区域允许一些外部路由通告到 OSPF 自治系统内部,而同时保留自治系统的区域部分的 STUB 区域的特征。假设一个 STUB 区域中的路由器连了一个运行其他路由进程的自治系统,现在这台路由器就变成了 ASBR,所以这个区域就不能再称为 STUB 区域了。如果把这个区域配置成一个 NSSA 区域,ASBR 会产生 NSSA 外部 LSA(类型 7),可以泛洪到整个 NSSA 区域。这些 7 类 LSA 在 NSSA ABR 上会转换成 5 类 LSA 并且泛洪到整个 OSPF 域中。

⑤完全 NSSA 区域。

和 NSSA 区域相似,完全 NSSA 区域允许一些外部路由通告到 OSPF 自治系统内部,而同时保留自治系统区域部分的完全 STUB 区域的特征。该区域的 ASBR 会产生 NSSA 外部 LSA(类型 7)在其区域内部泛洪并通过该区域的 ABR 转换成 5 类 LSA 在整个 OSPF 域泛洪。同时,该区域的 ABR 也会产生一条默认路由 0.0.0.0 传播到区域内,所有域间路由都必须通过 ABR 才能到达。

(2)默认路由的产生。

①普通区域。

默认情况下,在普通 OSPF 区域内的 OSPF 路由器是不会产生默认路由的,即使它有默认路由。

当网络中默认路由通过其他路由进程产生时,必须能够将默认路由通告到整个 OSPF 域中。这个时候要想产生默认路由必须在 ASBR 上 OSPF 协议视图下手动配置:

  [Router-ospf-1]default-route-advertise [always]

使用了该命令将在整个 OSPF 域中通告默认路由 0.0.0.0,但前提是该 ASBR 自己已经有默认路由,否则不会通告默认路由。

如果在该命令上加上关键字 always 的话,则无论 ASBR 是否有默认路由都将在整个 OSPF 域中通告默认路由 0.0.0.0,这将强制默认路由总是出现在路由表中,所以慎用关键字 always。

使用了该命令后将会产生一个链路状态 ID 为 0.0.0.0,网络掩码为 0.0.0.0 的 ASE LSA(5 类),并且通告到整个 OSPF 域中。

②STUB 区域。

由于 STUB 区域不允许外部 LSA 在其内部泛洪,所以该区域内的路由器除了 ABR 外没有自治系统外部路由,如果它们想到自治系统外部时应该怎么办?在 STUB 区域里的路由器将本区域内 ABR 作为出口,ABR 会产生默认路由 0.0.0.0 通告给整个 STUB 区域内的路由器,这样的话到达自治系统外部的路由可以通过 ABR 到达。

配置了 STUB 区域之后,ABR 自动会产生一条 Link ID 为 0.0.0.0,网络掩码为 0.0.0.0 的 SUMMARY LSA(3 类),并且通告到整个 STUB 区域内。

③完全 STUB 区域。

完全 STUB 区域不仅不允许外部 LSA 在其内部泛洪,连区域间的路由也不允许携带,所以在完全 STUB 区域里的路由器要想到别的区域或自治系统外部时应该怎么办呢?同样,在完全 STUB 区域里的路由器也将本区域内 ABR 作为出口,ABR 会产生默认路由 0.0.0.0 通告给整个完全 STUB 区域内的路由器,这样,到达本区域外部的路由都通过 ABR 到达就可以了。

配置了完全 STUB 区域之后,ABR 自动会产生一条 Link ID 为 0.0.0.0,网络掩码为 0.0.0.0 的 SUMMARY LSA(3 类),并且通告到整个完全 STUB 区域内。

④NSSA 区域。

NSSA 区域允许少量外部路由通过本区域的 ASBR 通告进来，它不允许携带其他区域的外部路由，这样的话到达自治系统外部路由只能通过本区域的 ASBR 到达，如果该 ASBR 没有通告该外路由的，则不能到达。

在只配置了 NSSA 区域的时候，是不会自动产生默认路由的。

⑤完全 NSSA 区域。

完全 NSSA 区域和 NSSA 区域不同的是，它不允许携带区域间路由，如果要到其他区域的时候应该怎么办呢？同样的，默认路由又出场了，在该区域 ABR 上会产生一条默认路由 0.0.0.0，通告给整个完全 NSSA 区域，所有的域间路由都将 NSSA ABR 作为出口。

配置完全 NSSA 区域后，就会自动产生一条 Link ID 为 0.0.0.0，网络掩码为 0.0.0.0 的 SUMMARY LSA（3 类），在 NSSA 区域内通告默认路由 0.0.0.0。

与 NSSA 区域 ABR 上默认路由产生的方式不同的是，在完全 NSSA 区域 ABR 上的默认路由是配置好区域之后自动产生类型 3 的默认 LSA，在 NSSA 区域上 ABR 的默认路由是自己可配置的，因为在完全 NSSA 区域产生的默认路由是必需的，它起着指导本区域内路由器区域间路由的作用。

14.3 项目实施

1. 项目拓扑图

OSPF 的缺省路由如图 14-1 所示。

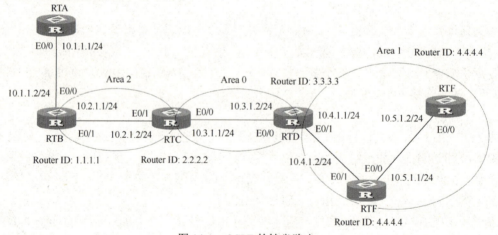

图 14-1 OSPF 的缺省路由

2. 项目实训环境准备

MSR20-40（6 台），计算机（1 台）、网线若干条。CMW 版本：5.20。

为了不受原来的配置影响，在实训之前先将所有的配置数据擦除后重新启动，命令为：

    <h3c>Reset saved-configuration
    <h3c>reboot

3. 项目主要实训步骤

（1）按照实训拓扑图连接设备，并给设备命名并设置 IP 地址。（略）

（2）给每台设备配置 Router ID 配置，并配置 OSPF 相关设置。（略）
（3）在 RTF 上显示 OSPF 的路由信息如下：

[RTF]dis ospf rou

OSPF Process 1 with Router ID 5.5.5.5
Routing Tables
Routing for Network

Destination	Cost	Type	NextHop	AdvRouter	Area
10.2.1.0/24	4	Inter	10.5.1.1	3.3.3.3	0.0.0.1
10.3.1.0/24	3	Inter	10.5.1.1	3.3.3.3	0.0.0.1
10.4.1.0/24	2	Transit	10.5.1.1	3.3.3.3	0.0.0.1
10.5.1.0/24	1	Transit	10.5.1.2	4.4.4.4	0.0.0.1

Total Nets: 4
Intra Area: 2   Inter Area: 2   ASE: 0   NSSA: 0

我们发现并没有达到 RTA（10.1.1.0 网段）的路由。默认情况下普通区域不产生默认路由。Area 1 和 Area 2 都是普通区域。RTB 是普通区域的 ASBR。

（4）在 RTB 上进行默认路由配置，并引入默认路由。

[RTB]ip route-static 0.0.0.0 0.0.0.0 10.1.1.1 preference 60
[RTB]ospf
[RTB-ospf-1]import-route static
[RTB-ospf-1]default-route-advertise
[RTB-ospf-1]quit
[RTB]dis ospf lsdb

OSPF Process 1 with Router ID 10.2.1.1
Link State Database
Area: 0.0.0.2

Type	LinkState ID	AdvRouter	Age	Len	Sequence	Metric
Router	10.2.1.1	10.2.1.1	20	36	80000002	0
Router	2.2.2.2	2.2.2.2	13	36	80000007	0
Network	10.2.1.2	2.2.2.2	17	32	80000002	0
Sum-Net	10.3.1.0	2.2.2.2	213	28	80000002	1
Sum-Net	10.4.1.0	2.2.2.2	213	28	80000002	2
Sum-Net	10.5.1.0	2.2.2.2	213	28	80000002	3

AS External Database

Type	LinkState ID	AdvRouter	Age	Len	Sequence	Metric
External	0.0.0.0	10.2.1.1	27	36	80000001	1

通过以上信息，可以发现 RTB 的链路状态信息库产生了一条默认 LSA，LinkState ID 为 0.0.0.0。

RTC 上的 lsdb 信息：

[RTC]dis ospf lsdb
OSPF Process 1 with Router ID 2.2.2.2
Link State Database

Area: 0.0.0.0

Type	LinkState ID	AdvRouter	Age	Len	Sequence	Metric
Router	3.3.3.3	3.3.3.3	531	36	80000004	0
Router	2.2.2.2	2.2.2.2	346	36	80000006	0
Network	10.3.1.1	2.2.2.2	346	32	80000003	0
Sum-Net	10.5.1.0	3.3.3.3	407	28	80000002	2
Sum-Net	10.2.1.0	2.2.2.2	346	28	80000002	1
Sum-Net	10.4.1.0	3.3.3.3	515	28	80000002	1
Sum-Asbr	10.2.1.1	2.2.2.2	151	28	80000001	1

Area: 0.0.0.2

Type	LinkState ID	AdvRouter	Age	Len	Sequence	Metric
Router	10.2.1.1	10.2.1.1	156	36	80000002	0
Router	2.2.2.2	2.2.2.2	146	36	80000007	0
Network	10.2.1.2	2.2.2.2	151	32	80000002	0
Sum-Net	10.5.1.0	2.2.2.2	347	28	80000002	3
Sum-Net	10.3.1.0	2.2.2.2	347	28	80000002	1
Sum-Net	10.4.1.0	2.2.2.2	347	28	80000002	2

AS External Database

Type	LinkState ID	AdvRouter	Age	Len	Sequence	Metric
External	0.0.0.0	10.2.1.1	162	36	80000001	1

RTD 上的 lsdb 信息：

```
[RTD]dis ospf lsdb
```

OSPF Process 1 with Router ID 3.3.3.3
Link State Database
Area: 0.0.0.0

Type	LinkState ID	AdvRouter	Age	Len	Sequence	Metric
Router	3.3.3.3	3.3.3.3	667	36	80000004	0
Router	2.2.2.2	2.2.2.2	485	36	80000006	0
Network	10.3.1.1	2.2.2.2	485	32	80000003	0
Sum-Net	10.4.1.0	3.3.3.3	651	28	80000002	1
Sum-Net	10.5.1.0	3.3.3.3	543	28	80000002	2
Sum-Net	10.2.1.0	2.2.2.2	485	28	80000002	1
Sum-Asbr	10.2.1.1	2.2.2.2	290	28	80000001	1

Area: 0.0.0.1

Type	LinkState ID	AdvRouter	Age	Len	Sequence	Metric
Router	5.5.5.5	5.5.5.5	455	36	80000003	0
Router	3.3.3.3	3.3.3.3	556	36	80000005	0
Router	4.4.4.4	4.4.4.4	453	48	80000007	0
Network	10.5.1.1	4.4.4.4	449	32	80000003	0
Network	10.4.1.1	3.3.3.3	552	32	80000003	0
Sum-Net	10.3.1.0	3.3.3.3	652	28	80000002	1
Sum-Net	10.2.1.0	3.3.3.3	652	28	80000002	2
Sum-Asbr	10.2.1.1	3.3.3.3	290	28	80000001	2

AS External Database

Type	LinkState ID	AdvRouter	Age	Len	Sequence	Metric
External	0.0.0.0	10.2.1.1	306	36	80000001	1

大家还可以观察其他路由的 LSDB 信息，都产生了默认路由，通告者是 RTB（AdvRouter：

10.2.1.1)。可以看出,普通区域的 ASBR 产生的默认路由 0.0.0.0 不仅仅在本区域内泛洪,还泛洪到整个 OSPF 区域。

14.4　项目总结与提高

写出主要项目实施规划、步骤与实训所得的主要结论。

项目十五 配置 OSPF 的 Stub 区域

15.1 项目提出

小王在某公司上班,想了解 OSPF 的 Stub 区域,于是设计了如下实验。

15.2 项目分析

1. 项目实训目的

理解 OSPF Stub 区域的功能。

2. 项目实现功能

通过实训,理解 OSPF Stub 区域的功能。

3. 项目主要应用的技术介绍

STUB 区域的设计思想在于:在划分了区域之后,非骨干区域中的路由器对于区域外的路由,一定要通过 ABR(区域边界路由器)来转发,或者说对于区域内的路由器来说 ABR 是一个通往外部世界的必经之路。既然如此,对于区域内的路由器来说,就没有必要知道通往外部世界的详细的路由了,代之以由 ABR 向该区域发布一条默认路由来指导报文的发送。这样在区域内的路由器中就只有为数不多的区域内路由和一条指向 ABR 的默认路由。而且无论区域外的路由如何变化,都不会影响到区域内路由器的路由表。由于区域内的路由器通常是由一些处理能力有限的低端路由器组成的,所以处于 STUB 区域内的这些低端设备既不需要保存庞大的路由表,也不需要经常进行路由计算。有了 STUB 属性之后,网络的规划更符合实际的设备特点。

以上描述的只是 STUB 区域的设计思想,在协议文本中,对 STUB 区域的精确定义是:STUB 区域一定是非骨干区域和非转换区域(可以配置虚连接的区域),并且在该区域中不可传递 Type 5 类型的 LSA。 因为协议的设计者认为路由表中的绝大部分路由均是来自自治系统外部的引入的路由。(由于 OSPF 是链路状态算法的路由协议,LSA 就是用来描述网络拓扑结构的一种数据结构。在 OSPF 中将 LSA 分为 5 类:type1、2 两种用来描述区域内的路由信息;type3 用来描述区域间的路由信息;type4、5 用来描述自治系统外部的路由信息)OSPF 链路状态公布 LSA 类型 5 定义了到达外部网络的路由,它并不泛洪到 STUB 区域。连接到 STUB 区域的 ABR 为外部网络发送一个默认路由(0.0.0.0)到 STUB 区域。这允许 STUB 区域内部的某个路由器将报文转发到一个目的网络,而该网络并没有出现在 STUB 区域路由器的路由表中。对于那些在自己的路由表中没有找到的网络报文,STUB 区域路由器将其转发到 ABR 路由器,而该 ABR 路由器已发送 0.0.0.0 LSA

需要注意的是定义中对于过滤 TYPE5 类型的 LSA 使用的描述语言是"不可传递"的,这就意味着不仅区域外的 ASE(自治系统外部)路由无法传递到 STUB 区域中,同时 STUB

区域内部的 ASE 路由也无法传递到本区域之外。换一句更通俗的话来描述：STUB 区域内的路由器都不可引入任何外部的路由（包括静态路由）。

STUB 区域对 LSA 的影响如下：

①从其他区域来的汇总 LSA 被引入。

②默认路由作为一个路由汇总被引入。

③外部 LSA 不被引入。

15.3 项目实施

1．项目拓扑图

配置 OSPF 的 STUB 区域如图 15-1 所示。

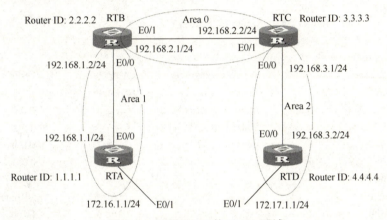

图 15-1　配置 OSPF 的 STUB 区域

2．项目实训环境准备

MSR20-40（4 台），计算机（1 台）、网线若干条。CMW 版本：5.20。

为了不受原来的配置影响，在实训之前先将所有的配置数据擦除后重新启动，命令为：

```
<h3c>Reset saved-configuration
<h3c>reboot
```

3．项目主要实训步骤

【实训要求】

（1）所有的路由器都运行 OSPF，整个自治系统划分为 3 个区域。

（2）其中 RouterA 和 RouterB 作为 ABR 来转发区域之间的路由，RouterD 作为 ASBR 引入了外部路由（静态路由）。

（3）要求将 Area1 配置为 STUB 区域，减少通告到此区域内的 LSA 数量，但不影响路由的可达性。

【实训步骤】

（1）按照实训拓扑图连接设备，并给设备命名并设置 IP 地址。（略）

（2）给每台设备配置 Router ID 配置，并配置 OSPF 相关设置。（略）

（3）配置 RTD 引入静态路由。

```
[RTD]ip route-static 100.0.0.0 8 null 0
[RTD]ospf 1
[RTD-ospf-1]import-route static
```

(4) 查看 RouterC 的 ABR/ASBR 信息。

```
[RTC]display ospf abr-asbr
        OSPF Process 1 with Router ID 3.3.3.3
              Routing Table to ABR and ASBR
 Type    Destination    Area        Cost   Nexthop        RtType
 Intra   4.4.4.4        0.0.0.2     1      192.168.3.2    ASBR
 Intra   2.2.2.2        0.0.0.0     1      192.168.2.1    ABR
```

(5) 查看 RTC 的 OSPF 路由表。

```
[RTC]display ospf routing
        OSPF Process 1 with Router ID 3.3.3.3
              Routing Tables
 Routing for Network
 Destination      Cost    Type    NextHop        AdvRouter      Area
 192.168.3.0/24   1       Transit 192.168.3.1    4.4.4.4        0.0.0.2
 172.16.1.0/24    3       Inter   192.168.2.1    2.2.2.2        0.0.0.0
 172.17.1.0/24    2       Stub    192.168.3.2    4.4.4.4        0.0.0.2
 192.168.1.0/24   2       Inter   192.168.2.1    2.2.2.2        0.0.0.0
 192.168.2.0/24   1       Transit 192.168.2.2    2.2.2.2        0.0.0.0
 Routing for ASEs
 Destination      Cost    Type    Tag     NextHop        AdvRouter
 100.0.0.0/8      1       Type2   1       192.168.3.2    4.4.4.4
 Total Nets: 6
 Intra Area: 3   Inter Area: 2   ASE: 1   NSSA: 0
```

当 RouterC 所在区域为普通区域时，可以看到路由表中存在 AS 外部的路由。

(6) 配置 Area1 为 Stub 区域。

①配置 RTA。

```
[RTA] ospf
[RTA-ospf-1] area 1
[RTA-ospf-1-area-0.0.0.1] stub
[RTA-ospf-1-area-0.0.0.1] quit
[RTA-ospf-1]quit
```

②配置 RTB。

```
[RTB] ospf
[RTB-ospf-1] stub-router
[RTB-ospf-1] area 1
[RTB-ospf-1-area-0.0.0.1] stub
[RTB-ospf-1-area-0.0.0.1] quit
[RTB-ospf-1] quit
```

③查看 RTA 的路由表。

```
[RTA]dis ospf rou
        OSPF Process 1 with Router ID 1.1.1.1
            Routing Tables
Routing for Network
Destination      Cost      Type      NextHop        AdvRouter      Area
0.0.0.0/0        2         Inter     192.168.1.2    2.2.2.2        0.0.0.1
192.168.3.0/24   3         Inter     192.168.1.2    2.2.2.2        0.0.0.1
172.16.1.0/24    1         Stub      172.16.1.1     1.1.1.1        0.0.0.1
172.17.1.0/24    4         Inter     192.168.1.2    2.2.2.2        0.0.0.1
192.168.1.0/24   1         Transit   192.168.1.1    1.1.1.1        0.0.0.1
192.168.2.0/24   2         Inter     192.168.1.2    2.2.2.2        0.0.0.1
Total Nets: 6
Intra Area: 2    Inter Area: 4     ASE: 0    NSSA: 0
```

说明：当把 RouterC 所在区域配置为 STUB 区域时，已经看不到 AS 外部的路由，取而代之的是一条缺省路由。

（7）配置禁止向 Stub 区域通告 Type3 LSA。

①在 RTB 上配置相关命令。

```
[RTB]ospf
[RTB-ospf-1]area 1
[RTB-ospf-1-area-0.0.0.1]stub no-summary
```

②查看 RTA 的路由表。

```
[RTA]dis ospf routing
        OSPF Process 1 with Router ID 1.1.1.1
            Routing Tables
Routing for Network
Destination      Cost      Type      NextHop        AdvRouter      Area
0.0.0.0/0        2         Inter     192.168.1.2    2.2.2.2        0.0.0.1
172.16.1.0/24    1         Stub      172.16.1.1     1.1.1.1        0.0.0.1
192.168.1.0/24   1         Transit   192.168.1.1    1.1.1.1        0.0.0.1
Total Nets: 3
Intra Area: 2    Inter Area: 1     ASE: 0    NSSA: 0
```

说明：禁止向 STUB 区域通告 Summary LSA 后，STUB 路由器的路由表项进一步减少，只保留了一条通往区域外部的缺省路由。

15.4 项目总结与提高

写出主要项目实施规划、步骤与实训所得的主要结论。

项目十六 配置 OSPF 的 NSSA 区域

16.1 项目提出

所有的路由器都运行 OSPF，整个自治系统划分为 3 个区域。其中 RTB 和 RTC 作为 ABR 来转发区域之间的路由，RTD 作为 ASBR 引入了外部路由（静态路由）。要求将 Area1 配置为 NSSA 区域，同时将 RTA 配置为 ASBR 引入外部路由（静态路由），且路由信息可正确地在 AS 内传播。

16.2 项目分析

1. 项目实训目的

理解 OSPF 的 NSSA 区域的功能和特点。

2. 项目实现功能

通过实训，理解 OSPF 的 NSSA 区域的功能和特点。

3. 项目主要应用的技术介绍

NSSA（"not-so-stubby" area）

自治系统外的 ASE 路由不可以进入到 NSSA 区域中，但是 NSSA 区域内的路由器引入的 ASE 路由可以在 NSSA 中传播并发送到区域之外。即：取消了 STUB 关于 ASE 的双向传播的限制（区域外的进不来，区域里的也出不去），改为单向限制（区域外的进不来，区域里的能出去）。

由于是作为 OSPF 标准协议的一种扩展属性，应尽量减少与不支持该属性的路由器协调工作时的冲突和兼容性问题。

为了解决 ASE 单向传递的问题，NSSA 中重新定义了一种 LSA——Type 7 类型的 LSA，作为区域内的路由器引入外部路由时使用,该类型的 LSA 除了类型标识与 Type 5 不相同之外，其他内容基本一样。这样区域内的路由器就可以通过 LSA 的类型来判断是否该路由来自本区域内。但由于 Type 7 类的 LSA 是新定义的，对于不支持 NSSA 属性的路由器无法识别，所以协议规定：在 NSSA 的 ABR 上将 NSSA 内部产生的 Type 7 类型的 LSA 转化为 Type 5 类型的 LSA 再发布出去，并同时更改 LSA 的发布者为 ABR 自己。这样 NSSA 区域外的路由器就可以完全不用支持该属性。

NSSA 对 LSA 的影响：

（1）类型 7LSA 在一个 NSSA 区域内携带外部信息。

（2）类型 7LSA 在 NSSA 的 ABR 上被转化为 5lsa。

（3）不允许外部 LSA。

（4）汇总 LSA 被引入。

16.3 项目实施

1. 项目拓扑图

配置 OSPF 的 NSSA 区域如图 16-1 所示。

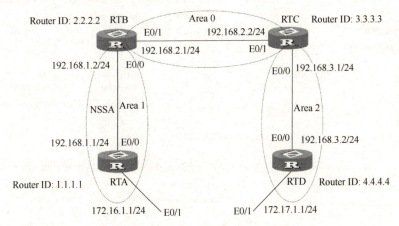

图 16-1 配置 OSPF 的 NSSA 区域

2. 项目实训环境准备

MSR20-40（4 台），计算机（1 台）、网线若干条。CMW 版本：5.20

为了不受原来的配置影响，在实训之前先将所有的配置数据擦除后重新启动，命令为：

    <h3c>Reset saved-configuration
    <h3c>reboot

3. 项目主要实训步骤

（1）按照实训拓扑图连接设备，并给设备命名并设置 IP 地址。（略）

（2）给每台设备配置 Router ID 配置，并配置 OSPF 相关设置。（略）

（3）配置 RTD 引入静态路由。

    [RTD]ip route-static 100.0.0.0 8 null 0
    [RTD]ospf 1
    [RTD-ospf-1]import-route static

（4）查看 RouterC 的 ABR/ASBR 信息。

    [RTC]display ospf abr-asbr

```
        OSPF Process 1 with Router ID 3.3.3.3
              Routing Table to ABR and ASBR
   Type    Destination    Area        Cost    Nexthop        RtType
   Intra   4.4.4.4        0.0.0.2     1       192.168.3.2    ASBR
   Intra   2.2.2.2        0.0.0.0     1       192.168.2.1    ABR
```

（5）查看 RTC 的 OSPF 路由表。

    [RTC]display ospf routing

```
        OSPF Process 1 with Router ID 3.3.3.3
                Routing Tables
Routing for Network
Destination       Cost    Type     NextHop        AdvRouter    Area
192.168.3.0/24    1       Transit  192.168.3.1    4.4.4.4      0.0.0.2
172.16.1.0/24     3       Inter    192.168.2.1    2.2.2.2      0.0.0.0
172.17.1.0/24     2       Stub     192.168.3.2    4.4.4.4      0.0.0.2
192.168.1.0/24    2       Inter    192.168.2.1    2.2.2.2      0.0.0.0
192.168.2.0/24    1       Transit  192.168.2.2    2.2.2.2      0.0.0.0
Routing for ASEs
Destination       Cost    Type     Tag            NextHop      AdvRouter
100.0.0.0/8       1       Type2    1              192.168.3.2  4.4.4.4
Total Nets: 6
Intra Area: 3    Inter Area: 2    ASE: 1    NSSA: 0
```

当 RouterC 所在区域为普通区域时，可以看到路由表中存在 AS 外部的路由。

（6）配置 Area1 为 NSSA 区域。

①配置 RTB。

```
[RTA] ospf
[RTA-ospf-1] area 1
[RTA-ospf-1-area-0.0.0.1] nssa default-route-advertise no-summary
[RTA-ospf-1-area-0.0.0.1]quit
[RTA-ospf-1]quit
```

②配置 RTA。

```
[RTB] ospf
[RTB-ospf-1] stub-router
[RTB-ospf-1] area 1
[RTB-ospf-1-area-0.0.0.1] nssa
[RTB-ospf-1-area-0.0.0.1] quit
[RTB-ospf-1] quit
```

注意：

建议在 ABR（这里的 RouterA）上配置 default-route-advertise no-summary 参数，这样可以减少 NSSA 路由器的路由表数量。

其他 NSSA 路由器只需配置 nssa 命令就可以。

③查看 RTA 的路由表。

```
[RTA]dis ospf rou
        OSPF Process 1 with Router ID 1.1.1.1
                Routing Tables
Routing for Network
Destination       Cost    Type     NextHop        AdvRouter    Area
0.0.0.0/0         2       Inter    192.168.1.2    2.2.2.2      0.0.0.1
172.16.1.0/24     1       Stub     172.16.1.1     1.1.1.1      0.0.0.1
192.168.1.0/24    1       Transit  192.168.1.1    1.1.1.1      0.0.0.1
Total Nets: 3
```

Intra Area: 2   Inter Area: 1   ASE: 0   NSSA: 0

说明：当把 RouterC 所在区域配置为 NSSA 区域时，已经看不到 AS 外部的路由，取而代之的是一条缺省路由。

（7）配置 RTA 引入静态路由。

[RTA]ip route-static 200.0.0.0 8 null 0
[RTA]ospf 1
[RTA-ospf-1]import-route st
[RTA-ospf-1]import-route static
[RTA-ospf-1]quit

（8）查看 RouterD 的 OSPF 路由表。

[RTD]dis ospf routing

OSPF Process 1 with Router ID 4.4.4.4
Routing Tables

Routing for Network

Destination	Cost	Type	NextHop	AdvRouter	Area
192.168.3.0/24	1	Transit	192.168.3.2	4.4.4.4	0.0.0.2
172.16.1.0/24	4	Inter	192.168.3.1	3.3.3.3	0.0.0.2
172.17.1.0/24	1	Stub	172.17.1.1	4.4.4.4	0.0.0.2
192.168.1.0/24	3	Inter	192.168.3.1	3.3.3.3	0.0.0.2
192.168.2.0/24	2	Inter	192.168.3.1	3.3.3.3	0.0.0.2

Routing for ASEs

Destination	Cost	Type	Tag	NextHop	AdvRouter
200.0.0.0/8	1	Type2	1	192.168.3.1	2.2.2.2

Total Nets: 6
Intra Area: 2   Inter Area: 3   ASE: 1   NSSA: 0

在 RTD 上可以看到 NSSA 区域引入的一条 AS 外部的路由。

16.4 项目总结与提高

写出主要项目实施规划、步骤与实训所得的主要结论。

项目十七 BGP 的基本配置

17.1 项目提出：

小王在某公司上班，想了解 BGP 协议，于是他设计了如下实验。

17.2 项目分析

1. 项目实训目的

掌握配置 BGP 的基本命令；
掌握 BGP 邻居关系的建立；
理解路由引入。

2. 项目实现功能

通过实训，掌握配置 BGP 的基本命令，掌握 BGP 邻居关系的建立，理解路由引入。

3. 项目主要应用的技术介绍

边界网关协议（BGP）是运行于 TCP 上的一种自治系统的路由协议。BGP 是唯一一个用来处理像因特网大小的网络的协议，也是唯一能够妥善处理好不相关路由域间的多路连接的协议。BGP 构建在 EGP 的经验之上。

BGP 系统的主要功能是和其他的 BGP 系统交换网络可达信息。网络可达信息包括列出的自治系统（AS）的信息。这些信息有效地构造了 AS 互联的拓扑图并由此清除了路由环路，同时在 AS 级别上可实施策略决策。

17.3 项目实施

1. 项目拓扑图

BGP 的基本配置如图 17-1 所示。

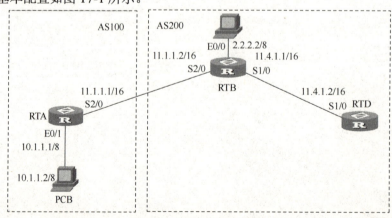

图 17-1 BGP 的基本配置

2. 项目实训环境准备

MSR20-40（3 台），计算机（2 台）、网线若干条。CMW 版本：5.20。

为了不受原来的配置影响，在实训之前先将所有的配置数据擦除后重新启动，命令为：

    \<h3c\>Reset saved-configuration
    \<h3c\>reboot

3. 项目主要实训步骤

（1）按照实训拓扑图连接设备，并给设备命名并设置 IP 地址。（略）

（2）配置 BGP。

①配置 RTA。

    [RTA]bgp 100
    [RTA-bgp]group as200 external
    [RTA-bgp]peer as200 as-number 200
    [RTA-bgp]peer 11.1.1.2 group as200
    [RTA-bgp]peer 11.2.1.2 group as200

②配置 RTB。

    [RTB]bgp 200
    [RTB-bgp]group as100 external
    [RTB-bgp]peer as100 as-number 100
    [RTB-bgp]peer 11.1.1.1 group as100

③配置 RTC。

    [RTC]bgp 200
    [RTC-bgp]group as100 external
    [RTC-bgp]peer as100 as-number 100
    [RTC-bgp]peer 11.2.1.1 group as100

（3）在 RTA 上查看 BGP 信息。

    [RTA]display bgp peer
     BGP local router ID : 11.2.1.1
     Local AS number : 100
     Total number of peers : 2   Peers in established state : 1
     Peer   AS   MsgRcvd   MsgSent OutQ PrefRcv Up/Down   State
     11.1.1.2   200   4   4   0   0   00:02:45 Established

对等体对应的 State 为"Estabilished"说明 RTA 与 RTB、RTC 的 EBGP 连接均已建立在 RTA 上执行：

    [RTA]dis bgp routing
     Total Number of Routes: 0

为什么没有任何 BGP 路由？因为 BGP 本身不会产生路由，只能发布引入的路由。

（4）在 RTA 上引入 10.0.0.0/8 网段的路由：

    [RTA]BGP 100

[RTA-bgp]network 10.0.0.0
[RTA-bgp]quit

再次查看路由表：

[RTA]dis bgp routing-table
 Total Number of Routes: 1
 BGP Local router ID is 11.2.1.1
 Status codes: * - valid, > - best, d - damped,
           h - history, i - internal, s - suppressed, S - Stale
      Origin : i - IGP, e - EGP, ? - incomplete

	Network	NextHop	MED	LocPrf	PrefVal Path/Ogn
*>	10.0.0.0	0.0.0.0	0		0 i

在 RTB 上查看路由表：

[RTB]dis ip routing
Routing Tables: Public
        Destinations : 8   Routes : 8

Destination/Mask	Proto	Pre	Cost	NextHop	Interface
2.0.0.0/8	Direct	0	0	2.2.2.2	Eth0/0
2.2.2.2/32	Direct	0	0	127.0.0.1	InLoop0
10.0.0.0/8	BGP	255	0	11.1.1.1	S2/0
11.1.0.0/16	Direct	0	0	11.1.1.2	S2/0
11.1.1.1/32	Direct	0	0	11.1.1.1	S2/0
11.1.1.2/32	Direct	0	0	127.0.0.1	InLoop0
127.0.0.0/8	Direct	0	0	127.0.0.1	InLoop0
127.0.0.1/32	Direct	0	0	127.0.0.1	InLoop0

[RTB]ping 10.1.1.1
  PING 10.1.1.1: 56   data bytes, press CTRL_C to break
    Reply from 10.1.1.1: bytes=56 Sequence=1 ttl=255 time=26 ms
    Reply from 10.1.1.1: bytes=56 Sequence=2 ttl=255 time=26 ms
    Reply from 10.1.1.1: bytes=56 Sequence=3 ttl=255 time=27 ms
    Reply from 10.1.1.1: bytes=56 Sequence=4 ttl=255 time=26 ms
    Reply from 10.1.1.1: bytes=56 Sequence=5 ttl=255 time=26 ms
  --- 10.1.1.1 ping statistics ---
   5 packet(s) transmitted
   5 packet(s) received
   0.00% packet loss
   round-trip min/avg/max = 26/26/27 ms

17.4 项目总结与提高

写出主要项目实施规划、步骤与实训所得的主要结论。

项目十八　BGP 的路由聚合

18.1　项目提出

小王在某公司上班,想了解 BGP 的路由聚合的配置方法,于是他设计了如下实验。

18.2　项目分析

1．项目实训目的

理解 BGP 路由引入和路由通告原则;
掌握 BGP 路由聚合的配置方法。

2．项目实现功能

通过实训,理解 BGP 路由引入和路由通告原则,掌握 BGP 路由聚合的配置方法。

3．项目主要应用的技术介绍

(1) 路由聚合的必要性。

在大规模的网络中,BGP 路由表变得十分庞大,存储路由表占有大量的路由器内存资源,传输和处理路由信息所必需的带宽和路由器传送与处理路由信息需要大量的资源。使用路由聚合（Routes Aggregation）可以大大减小路由表的规模;另外通过对路由的条目的聚合,隐藏一些具体的路由从而减少路由震荡对网络带来的影响。BGP 路由聚合结合灵活的路由策略,从而使 BGP 更有效地传递和控制路由。

(2) 路由聚合的方法。

聚合路由的方式:通过与静态路由组合对具体的路由条目进行路由聚合,自动聚合,手动聚合。

18.3　项目实施

1．项目拓扑图

BGP 的路由聚合如图 18-1 所示。

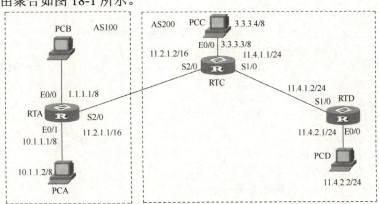

图 18-1　BGP 的路由聚合

2. 项目实训环境准备

MSR20-40（3 台），计算机（4 台）、网线若干条。CMW 版本：5.20。

为了不受原来的配置影响，在实训之前先将所有的配置数据擦除后重新启动，命令为：

<h3c>Reset saved-configuration
<h3c>reboot

3. 项目主要实训步骤

（1）按照实训拓扑图连接设备，并给设备命名并设置 IP 地址。（略）

（2）配置 BGP。

①配置 RTA。

[RTA]bgp 100
[RTA-bgp]group as200 external
[RTA-bgp]peer as200 as-number 200
[RTA-bgp]peer 11.2.1.2 group as200

②配置 RTC。

[RTC]bgp 200
[RTC-bgp]group as100 external
[RTC-bgp]peer as100 as-number 100
[RTC-bgp]peer 11.2.1.1 group as100
[RTC-bgp]group as200
[RTC-bgp]peer 11.4.1.2 group as200

（3）在 RTA 上引入 10.0.0.0/8 网段的路由：

[RTA]BGP 100
[RTA-bgp]network 10.0.0.0
[RTA-bgp]quit

再次查看路由表：

[RTA]dis bgp routing-table
 Total Number of Routes: 1
 BGP Local router ID is 11.2.1.1
 Status codes: * - valid, > - best, d - damped,
               h - history, i - internal, s - suppressed, S - Stale
       Origin : i - IGP, e - EGP, ? - incomplete

Network	NextHop	MED	LocPrf	PrefVal	Path/Ogn
*> 10.0.0.0	0.0.0.0	0		0	i

（4）查看 RTD 路由表：

[RTD]dis bgp routing
 Total Number of Routes: 1
 BGP Local router ID is 11.4.2.1
 Status codes: * - valid, > - best, d - damped,
               h - history,   i - internal, s - suppressed, S - Stale

项目十八 BGP 的路由聚合

Origin : i - IGP, e - EGP, ? - incomplete

Network	NextHop	MED	LocPrf	PrefVal	Path/Ogn
i 10.0.0.0	11.2.1.1	0	100	0	100i

[RTD]dis ip rou
Routing Tables: Public
    Destinations : 7   Routes : 7

Destination/Mask	Proto	Pre	Cost	NextHop	Interface
11.4.1.0/24	Direct	0	0	11.4.1.2	S1/0
11.4.1.1/32	Direct	0	0	11.4.1.1	S1/0
11.4.1.2/32	Direct	0	0	127.0.0.1	InLoop0
11.4.2.0/24	Direct	0	0	11.4.2.1	Eth0/0
11.4.2.1/32	Direct	0	0	127.0.0.1	InLoop0
127.0.0.0/8	Direct	0	0	127.0.0.1	InLoop0
127.0.0.1/32	Direct	0	0	127.0.0.1	InLoop0

为什么在 RTD 的路由表中没有 10.0.0.0/8 网段的路由呢？

根据 BGP 的选路原则，RTD 因为无法达到通往 10.0.0.0/8 网段的下一跳 11.2.1.1 而丢弃该路由。修改措施之一，在 RTC 处修改路由下一跳属性，将其改为与 RTD 建立 IBGP 连接的接口地址。

[RTC-bgp]peer as200 next-hop-local
[RTC-bgp]dis ip routing
Routing Tables: Public
    Destinations : 11   Routes : 11

Destination/Mask	Proto	Pre	Cost	NextHop	Interface
3.0.0.0/8	Direct	0	0	3.3.3.3	Eth0/0
3.3.3.3/32	Direct	0	0	127.0.0.1	InLoop0
10.0.0.0/8	BGP	255	0	11.2.1.1	S2/0
11.2.0.0/16	Direct	0	0	11.2.1.2	S2/0
11.2.1.1/32	Direct	0	0	11.2.1.1	S2/0
11.2.1.2/32	Direct	0	0	127.0.0.1	InLoop0
11.4.1.0/24	Direct	0	0	11.4.1.1	S1/0
11.4.1.1/32	Direct	0	0	127.0.0.1	InLoop0
11.4.1.2/32	Direct	0	0	11.4.1.2	S1/0
127.0.0.0/8	Direct	0	0	127.0.0.1	InLoop0
127.0.0.1/32	Direct	0	0	127.0.0.1	InLoop0

（5）引入直连路由。

[RTD]ping 10.1.1.1
 PING 10.1.1.1: 56 data bytes, press CTRL_C to break
  Request time out
  Request time out
  Request time out
  Request time out
  Request time out
 --- 10.1.1.1 ping statistics ---
 5 packet(s) transmitted

0 packet(s) received
100.00% packet loss

RTD 仍然不能 ping 通 RTA，什么原因呢？

查看 RTA 的路由表。

[RTA]dis ip rou
Routing Tables: Public
        Destinations : 9   Routes : 9

Destination/Mask	Proto	Pre	Cost	NextHop	Interface
1.0.0.0/8	Direct	0	0	1.1.1.1	Eth0/0
1.1.1.1/32	Direct	0	0	127.0.0.1	InLoop0
10.0.0.0/8	Direct	0	0	10.1.1.1	Eth0/1
10.1.1.1/32	Direct	0	0	127.0.0.1	InLoop0
11.2.0.0/16	Direct	0	0	11.2.1.1	S2/0
11.2.1.1/32	Direct	0	0	127.0.0.1	InLoop0
11.2.1.2/32	Direct	0	0	11.2.1.2	S2/0
127.0.0.0/8	Direct	0	0	127.0.0.1	InLoop0
127.0.0.1/32	Direct	0	0	127.0.0.1	InLoop0

原来 RTA 没有到 RTD 的路由。

将 RTD 的直连路由引入 BGP：

[RTD-BGP]import-route direct
[RTA]dis bgp routing
 Total Number of Routes: 2
 BGP Local router ID is 11.2.1.1
 Status codes: * - valid, > - best, d - damped,
               h - history,   i - internal, s - suppressed, S - Stale
               Origin : i - IGP, e - EGP, ? - incomplete

	Network	NextHop	MED	LocPrf	PrefVal	Path/Ogn
*>	10.0.0.0	0.0.0.0	0		0	i
*>	11.4.2.0/24	11.2.1.2			0	200

RTA 仍未学到 RTC 和 RTD 之间的网段 11.4.1.0/24。

[RTD]dis bgp routing
 Total Number of Routes: 6
 BGP Local router ID is 11.4.2.1
 Status codes: * - valid, > - best, d - damped,
               h - history,   i - internal, s - suppressed, S - Stale
               Origin : i - IGP, e - EGP, ? - incomplete

	Network	NextHop	MED	LocPrf	PrefVal	Path/Ogn
*>i	10.0.0.0	11.4.1.1	0	100	0	100i
*>	11.4.1.0/24	0.0.0.0	0		0	?
*>	11.4.1.1/32	0.0.0.0	0		0	?
*>	11.4.1.2/32	0.0.0.0	0		0	?
*>	11.4.2.0/24	0.0.0.0	0		0	?
*>	11.4.2.1/32	0.0.0.0	0		0	?

RTD 把路由传给了 RTC，但 RTC 没有传给 RTA，为什么呢？
原来 BGP Speaker RTC 发现自己已经有了直连路由：

```
 *>   11.4.1.0/24           0.0.0.0          0                         0          ?
```

而从 BGP Spesker RTD 获得的 BGP 路由：

```
 *>i  10.0.0.0              11.4.1.1         0          100            0          100i
```

不是最佳路由，RTC 不会采用这条 BGP 路由。由于 BGP 只选最优的路由给自己使用，也只把自己使用的路由通告给对等体，所以这条 BGP 路由没有被继续通告给 RTA。

```
[RTC]dis bgp routing
 Total Number of Routes: 4
 BGP Local router ID is 11.4.1.1
 Status codes: * - valid, > - best, d - damped,
               h - history,  i - internal, s - suppressed, S - Stale
               Origin : i - IGP, e - EGP, ? - incomplete

    Network            NextHop          MED        LocPrf         PrefVal    Path/Ogn
 *>   10.0.0.0         11.2.1.1         0                         0          100i
 * i  11.4.1.0/24      11.4.1.2         0          100            0          ?
 * i  11.4.1.1/32      11.4.1.2         0          100            0          ?
 *>i  11.4.2.0/24      11.4.1.2         0          100            0          ?
[RTC]dis ip rou
Routing Tables: Public
         Destinations : 12        Routes : 12
Destination/Mask     Proto    Pre    Cost        NextHop        Interface
3.0.0.0/8            Direct   0      0           3.3.3.3        Eth0/0
3.3.3.3/32           Direct   0      0           127.0.0.1      InLoop0
10.0.0.0/8           BGP      255    0           11.2.1.1       S2/0
11.2.0.0/16          Direct   0      0           11.2.1.2       S2/0
11.2.1.1/32          Direct   0      0           11.2.1.1       S2/0
11.2.1.2/32          Direct   0      0           127.0.0.1      InLoop0
11.4.1.0/24          Direct   0      0           11.4.1.1       S1/0
11.4.1.1/32          Direct   0      0           127.0.0.1      InLoop0
11.4.1.2/32          Direct   0      0           11.4.1.2       S1/0
11.4.2.0/24          BGP      255    0           11.4.1.2       S1/0
127.0.0.0/8          Direct   0      0           127.0.0.1      InLoop0
127.0.0.1/32         Direct   0      0           127.0.0.1      InLoop0
```

如何解决呢？方法就是 RTC 将自己的直连路由映入 BGP。

```
[RTC]
[RTC]BGP 200
[RTC-bgp]import-route direct
[RTC-bgp]dis bgp routing
 Total Number of Routes: 12
 BGP Local router ID is 11.4.1.1
 Status codes: * - valid, > - best, d - damped,
               h - history,  i - internal, s - suppressed, S - Stale
               Origin : i - IGP, e - EGP, ? - incomplete

    Network            NextHop          MED        LocPrf         PrefVal    Path/Ogn
 *>   3.0.0.0          0.0.0.0          0                         0          ?
```

*>	3.3.3.3/32	0.0.0.0	0		0	?
*>	10.0.0.0	11.2.1.1	0		0	100i
*>	11.2.0.0/16	0.0.0.0	0		0	?
*>	11.2.1.1/32	0.0.0.0	0		0	?
*>	11.2.1.2/32	0.0.0.0	0		0	?
*>	11.4.1.0/24	0.0.0.0	0		0	?
* i		11.4.1.2	0	100	0	?
*>	11.4.1.1/32	0.0.0.0	0		0	?
* i		11.4.1.2	0	100	0	?
*>	11.4.1.2/32	0.0.0.0	0		0	?
*>i	11.4.2.0/24	11.4.1.2	0	100	0	?

在 RTA 上查看 BGP 路由信息：

[RTA]display bgp routing
Total Number of Routes: 7
BGP Local router ID is 11.2.1.1
Status codes: * - valid, > - best, d - damped,
    h - history, i - internal, s - suppressed, S - Stale
  Origin : i - IGP, e - EGP, ? - incomplete

	Network	NextHop	MED	LocPrf	PrefVal Path/Ogn
*>	3.0.0.0	11.2.1.2	0	0	200?
*>	10.0.0.0	0.0.0.0	0	0	i
*	11.2.0.0/16	11.2.1.2	0	0	200?
*	11.2.1.1/32	11.2.1.2	0	0	200?
*>	11.4.1.0/24	11.2.1.2	0	0	200?
*>	11.4.1.2/32	11.2.1.2	0	0	200?
*>	11.4.2.0/24	11.2.1.2		0	200?

[RTA]display ip routing
Routing Tables: Public
  Destinations : 13 Routes : 13

Destination/Mask	Proto	Pre	Cost	NextHop	Interface
1.0.0.0/8	Direct	0	0	1.1.1.1	Eth0/0
1.1.1.1/32	Direct	0	0	127.0.0.1	InLoop0
3.0.0.0/8	BGP	255	0	11.2.1.2	S2/0
10.0.0.0/8	Direct	0	0	10.1.1.1	Eth0/1
10.1.1.1/32	Direct	0	0	127.0.0.1	InLoop0
11.2.0.0/16	Direct	0	0	11.2.1.1	S2/0
11.2.1.1/32	Direct	0	0	127.0.0.1	InLoop0
11.2.1.2/32	Direct	0	0	11.2.1.2	S2/0
11.4.1.0/24	BGP	255	0	11.2.1.2	S2/0
11.4.1.2/32	BGP	255	0	11.2.1.2	S2/0
11.4.2.0/24	BGP	255	0	11.2.1.2	S2/0
127.0.0.0/8	Direct	0	0	127.0.0.1	InLoop0
127.0.0.1/32	Direct	0	0	127.0.0.1	InLoop0

在 RTD 上 ping 10.1.1.1，显示结果如下：

[RTD]ping 10.1.1.1

项目十八 BGP 的路由聚合

PING 10.1.1.1: 56   data bytes, press CTRL_C to break
 Reply from 10.1.1.1: bytes=56 Sequence=1 ttl=254 time=52 ms
 Reply from 10.1.1.1: bytes=56 Sequence=2 ttl=254 time=51 ms
 Reply from 10.1.1.1: bytes=56 Sequence=3 ttl=254 time=51 ms
 Reply from 10.1.1.1: bytes=56 Sequence=4 ttl=254 time=51 ms
 Reply from 10.1.1.1: bytes=56 Sequence=5 ttl=254 time=51 ms
--- 10.1.1.1 ping statistics ---
 5 packet(s) transmitted
 5 packet(s) received
 0.00% packet loss
 round-trip min/avg/max = 51/51/52 ms

（6）配置路由聚合。

[RTA]dis ip routing
Routing Tables: Public
   Destinations : 13 Routes : 13

Destination/Mask	Proto	Pre	Cost	NextHop	Interface
1.0.0.0/8	Direct	0	0	1.1.1.1	Eth0/0
1.1.1.1/32	Direct	0	0	127.0.0.1	InLoop0
3.0.0.0/8	BGP	255	0	11.2.1.2	S2/0
10.0.0.0/8	Direct	0	0	10.1.1.1	Eth0/1
10.1.1.1/32	Direct	0	0	127.0.0.1	InLoop0
11.2.0.0/16	Direct	0	0	11.2.1.1	S2/0
11.2.1.1/32	Direct	0	0	127.0.0.1	InLoop0
11.2.1.2/32	Direct	0	0	11.2.1.2	S2/0
11.4.1.0/24	BGP	255	0	11.2.1.2	S2/0
11.4.1.2/32	BGP	255	0	11.2.1.2	S2/0
11.4.2.0/24	BGP	255	0	11.2.1.2	S2/0
127.0.0.0/8	Direct	0	0	127.0.0.1	InLoop0
127.0.0.1/32	Direct	0	0	127.0.0.1	InLoop0

将 BGP Speaker RTC 传来的两条路由 11.4.1.0/24 和 11.4.2.0/24 进行路由聚合。

[RTC-bgp]aggregate 11.4.0.0 255.255.0.0 detail-suppressed
[RTC-bgp]display BGP routing
Total Number of Routes: 13
BGP Local router ID is 11.4.1.1
Status codes: * - valid, > - best, d - damped,
    h - history,   i - internal, s - suppressed, S - Stale
   Origin : i - IGP, e - EGP, ? - incomplete

	Network	NextHop	MED	LocPrf	PrefVal	Path/Ogn
*>	3.0.0.0	0.0.0.0	0		0	?
*>	3.3.3.3/32	0.0.0.0	0		0	?
*>	10.0.0.0	11.2.1.1	0		0	100i
*>	11.2.0.0/16	0.0.0.0	0		0	?
*>	11.2.1.1/32	0.0.0.0	0		0	?
*>	11.2.1.2/32	0.0.0.0	0		0	?
*>	11.4.0.0/16	127.0.0.1			0	?
s>	11.4.1.0/24	0.0.0.0	0		0	?

* i		11.4.1.2	0	100	0	?
*>	11.4.1.1/32	0.0.0.0	0		0	?
* i		11.4.1.2	0	100	0	?
s>	11.4.1.2/32	0.0.0.0	0		0	?
s>	11.4.2.0/24	11.4.1.2	0	100	0	?

[RTC-bgp]
[RTA]dis bgp routing
  Total Number of Routes: 5
  BGP Local router ID is 11.2.1.1
  Status codes: * - valid, > - best, d - damped,
                h - history, i - internal, s - suppressed, S - Stale
                Origin : i - IGP, e - EGP, ? – incomplete

	Network	NextHop	MED	LocPrf	PrefVal	Path/Ogn
*>	3.0.0.0	11.2.1.2	0		0	200?
*>	10.0.0.0	0.0.0.0	0		0	i
*	11.2.0.0/16	11.2.1.2	0		0	200?
*	11.2.1.1/32	11.2.1.2	0		0	200?
*>	11.4.0.0/16	11.2.1.2			0	200?

[RTA]display ip routing
Routing Tables: Public
           Destinations : 11   Routes : 11

Destination/Mask	Proto	Pre	Cost	NextHop	Interface
1.0.0.0/8	Direct	0	0	1.1.1.1	Eth0/0
1.1.1.1/32	Direct	0	0	127.0.0.1	InLoop0
3.0.0.0/8	BGP	255	0	11.2.1.2	S2/0
10.0.0.0/8	Direct	0	0	10.1.1.1	Eth0/1
10.1.1.1/32	Direct	0	0	127.0.0.1	InLoop0
11.2.0.0/16	Direct	0	0	11.2.1.1	S2/0
11.2.1.1/32	Direct	0	0	127.0.0.1	InLoop0
11.2.1.2/32	Direct	0	0	11.2.1.2	S2/0
11.4.0.0/16	BGP	255	0	11.2.1.2	S2/0
127.0.0.0/8	Direct	0	0	127.0.0.1	InLoop0
127.0.0.1/32	Direct	0	0	127.0.0.1	InLoop0

[RTA]ping 11.4.2.1
  PING 11.4.2.1: 56   data bytes, press CTRL_C to break
    Reply from 11.4.2.1: bytes=56 Sequence=1 ttl=254 time=52 ms
    Reply from 11.4.2.1: bytes=56 Sequence=2 ttl=254 time=51 ms
    Reply from 11.4.2.1: bytes=56 Sequence=3 ttl=254 time=51 ms
    Reply from 11.4.2.1: bytes=56 Sequence=4 ttl=254 time=52 ms
    Reply from 11.4.2.1: bytes=56 Sequence=5 ttl=254 time=52 ms
  --- 11.4.2.1 ping statistics ---
    5 packet(s) transmitted
    5 packet(s) received
    0.00% packet loss
    round-trip min/avg/max = 51/51/52 ms

18.4 项目总结与提高

写出主要项目实施规划、步骤与实训所得的主要结论。

项目十九　BGP LOCAL-PREF 与 MED 属性的应用

19.1　项目提出

小王在某公司上班，想了解 BGP 协议 LOCAL-PREF 与 MED 属性的应用，于是他设计了如下实验。

19.2　项目分析

1．项目实训目的

掌握 BGP 路由调试的方法；

理解 local_preference 属性和 MED 属性。

2．项目实现功能

通过实训，掌握 BGP 路由调试的方法，理解 local_preference 属性和 MED 属性。

3．项目主要应用的技术介绍

default local-preference 命令用来配置本地优先级，使用 undo default local-preference 命令用来恢复本地优先级的缺省值。可以用配置不同本地优先级的方法来影响 BGP 的路由选择。

MED：multi_exit_disc（多出口鉴别器）。是在 eBGP 邻居间传递的属性，多出口区分（MED）属性是一个路由的外部度量（metric），与本地优先级属性不同，MED 在自治系统间交换，但进入 AS 的 MED 不离开该 AS。MED 属性用于选择最佳路由，MED 较小的路由被选择。当一个运行 BGP 的路由器通过不同的外部伙伴（External Peer）得到目的地址相同、下一跳不同的路由时，将根据不同路由的 MED 值进行优先选择。在其他条件相同的情况下，MED 较小的路由作为自治系统的外部路由。

19.3　项目实施

1．项目拓扑图

BGP LOCAL-PREF 与 MED 属性的应用如图 19-1 所示。

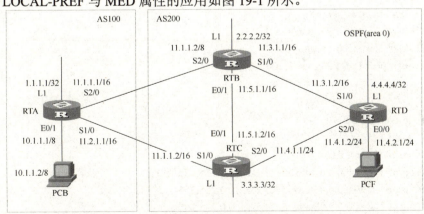

图 19-1　BGP LOCAL-PREF 与 MED 属性的应用

2. 项目实训环境准备

MSR20-40（4 台），计算机（2 台）、网线若干条。CMW 版本：5.20。

为了不受原来的配置影响，在实训之前先将所有的配置数据擦除后重新启动，命令为：

<h3c>Reset saved-configuration
<h3c>reboot

3. 项目主要实训步骤

（1）按照实训拓扑图连接设备，并给设备命名并设置 IP 地址。（略）

（2）在各路由器配置 BGP。

①RTA 的配置。

[RTA]BGP 100
[RTA-bgp]network 10.0.0.0
[RTA-bgp]import-route direct
[RTA-bgp]group as200 external
[RTA-bgp]peer as200 as-number 200
[RTA-bgp]peer 11.1.1.2 group as200
[RTA-bgp]peer 11.2.1.2 group as200

②RTB 的配置。

[RTB]router id 2.2.2.2
[RTB]ospf 1
[RTB-ospf-1]area 0.0.0.0
[RTB-ospf-1-area-0.0.0.0] network 11.3.0.0 0.0.255.255
[RTB-ospf-1-area-0.0.0.0]network 11.1.0.0 0.0.255.255
[RTB-ospf-1-area-0.0.0.0]network 11.5.0.0 0.0.255.255
[RTB-ospf-1-area-0.0.0.0]network 2.2.2.2 0.0.0.0
[RTB]bgp 200
[RTB-bgp]aggregate 11.4.0.0 255.255.0.0 detail-suppressed
[RTB-bgp]import-route ospf
[RTB-bgp]group as100 external
[RTB-bgp]peer as100 as-number 100
[RTB-bgp]peer 11.1.1.1 group as100
[RTB-bgp]group as200
[RTB-bgp]peer 11.3.1.2 group as200
[RTB-bgp]peer as200 next-hop-local
[RTB-bgp]peer 11.5.1.2 group as200

③RTC 的配置。

[RTC]router id 3.3.3.3
[RTC]ospf 1
[RTC-ospf-1]area 0.0.0.0
[RTC-ospf-1-area-0.0.0.0]network 11.2.1.2 0.0.255.255
[RTC-ospf-1-area-0.0.0.0]network 11.4.1.1 0.0.255.255
[RTC-ospf-1-area-0.0.0.0]network 11.5.1.2 0.0.255.255
[RTC-ospf-1-area-0.0.0.0]network 3.3.3.3 0.0.0.0

```
[RTC]bgp 200
[RTC-bgp]aggregate 11.4.0.0 255.255.0.0 detail-suppressed
[RTC-bgp]import-route ospf
[RTC-bgp]group as 100 external
[RTC-bgp]group as100 external
[RTC-bgp]peer as100 as-number 100
[RTC-bgp]group as200
[RTC-bgp]peer 11.2.1.1 group as100
[RTC-bgp]peer 11.4.1.2 group as200
[RTC-bgp]peer 11.5.1.1 group as200
[RTC-bgp]peer as200 next-hop-local
```

④RTD 的配置。

```
[RTD]router id 4.4.4.4
[RTD]ospf 1
[RTD-ospf-1]area 0.0.0.0
[RTD-ospf-1-area-0.0.0.0]network 11.3.1.2 0.0.255.255
%Aug  7 08:08:40:43 2010 RTD RM/3/RMLOG:OSPF-NBRCHANGE: Process 1, Neighbor 11.3
.1.1(Serial1/0) from Loading to Full
[RTD-ospf-1-area-0.0.0.0]network 11.4.2.1 0.0.0.255
[RTD-ospf-1-area-0.0.0.0]network 11.4.1.2 0.0.0.255
[RTD-ospf-1]quit
[RTD]bgp 200
[RTD-bgp]import-route ospf
[RTD-bgp]group as200
[RTD-bgp]peer 11.4.1.1 group as200
[RTD-bgp]peer 11.3.1.1 group as200
```

(3) 确认连接状态。

```
[RTB]display bgp peer
 BGP local router ID : 2.2.2.2
 Local AS number : 200
 Total number of peers : 3           Peers in established state : 3
   Peer            AS    MsgRcvd  MsgSent OutQ PrefRcv Up/Down    State
   11.1.1.1        100     29       33     0     6   00:28:50  Established
   11.3.1.2        200      8        8     0     5   00:01:11  Established
   11.5.1.2        200     21       19     0     6   00:13:40  Established
[RTC]dis bgp peer
 BGP local router ID : 3.3.3.3
 Local AS number : 200
 Total number of peers : 3           Peers in established state : 3
   Peer            AS    MsgRcvd  MsgSent OutQ PrefRcv Up/Down    State
   11.2.1.1        100      6        5     0     6   00:00:52  Established
   11.4.1.2        200      7       10     0     5   00:00:31  Established
   11.5.1.1        200      7        6     0     6   00:00:11  Established
```

(4) 利用属性改变从 RTD 至 RTA 的路由(local-preference 和 MED 属性)。

```
[RTD]dis ip routing
Routing Tables: Public
         Destinations : 21       Routes : 22
Destination/Mask    Proto   Pre   Cost      NextHop        Interface
1.1.1.1/32          BGP     255   0         11.3.1.1       S1/0
2.2.2.2/32          OSPF    10    1562      11.3.1.1       S1/0
3.3.3.3/32          OSPF    10    1562      11.4.1.1       S2/0
4.4.4.4/32          Direct  0     0         127.0.0.1      InLoop0
10.0.0.0/8          BGP     255   0         11.3.1.1       S1/0
10.4.2.0/24         Direct  0     0         10.4.2.1       Eth0/1
10.4.2.1/32         Direct  0     0         127.0.0.1      InLoop0
11.1.0.0/16         OSPF    10    3124      11.3.1.1       S1/0
11.1.1.2/32         BGP     255   0         11.4.1.1       S2/0
11.2.0.0/16         OSPF    10    3124      11.4.1.1       S2/0
11.2.1.2/32         BGP     255   0         11.3.1.1       S1/0
11.3.0.0/16         Direct  0     0         11.3.1.2       S1/0
11.3.1.1/32         Direct  0     0         11.3.1.1       S1/0
11.3.1.2/32         Direct  0     0         127.0.0.1      InLoop0
11.4.0.0/16         BGP     255   0         11.3.1.1       S1/0
11.4.1.0/24         Direct  0     0         11.4.1.2       S2/0
11.4.1.1/32         Direct  0     0         11.4.1.1       S2/0
11.4.1.2/32         Direct  0     0         127.0.0.1      InLoop0
11.5.0.0/16         OSPF    10    1563      11.3.1.1       S1/0
                    OSPF    10    1563      11.4.1.1       S2/0
127.0.0.0/8         Direct  0     0         127.0.0.1      InLoop0
127.0.0.1/32        Direct  0     0         127.0.0.1      InLoop0
[RTD]dis bgp routing
Total Number of Routes: 18
BGP Local router ID is 4.4.4.4
Status codes: * - valid, > - best, d - damped,
              h - history,  i - internal, s - suppressed, S - Stale
              Origin : i - IGP, e - EGP, ? - incomplete
       Network          NextHop       MED      LocPrf      PrefVal Path/Ogn
*>i 1.1.1.1/32          11.3.1.1      0        100         0       100?
*  i                    11.4.1.1      0        100         0       100?
*>   2.2.2.2/32         0.0.0.0       1562                 0       ?
*  i                    11.4.1.1      1        100         0       ?
*>   3.3.3.3/32         0.0.0.0       1562                 0       ?
*  i                    11.3.1.1      1        100         0       ?
*>i 10.0.0.0            11.3.1.1      0        100         0       100i
*  i                    11.4.1.1      0        100         0       100i
*>   11.1.0.0/16        0.0.0.0       3124                 0       ?
*  i                    11.4.1.1      1563     100         0       ?
*>i 11.1.1.2/32         11.4.1.1      0        100         0       100?
*>   11.2.0.0/16        0.0.0.0       3124                 0       ?
*  i                    11.3.1.1      1563     100         0       ?
*>i 11.2.1.2/32         11.3.1.1      0        100         0       100?
```

```
   * i  11.3.0.0/16         11.4.1.1      1563      100      0      ?
   *>i  11.4.0.0/16         11.3.1.1                100      0      ?
   *>   11.5.0.0/16         0.0.0.0       1563               0      ?
   *>                       0.0.0.0       1563               0      ?
```

从上面的 BGP 路由表可以看出,有两条从 RTD 到 RTA 的路由。如何改变去 RTA 的路由,经过 RTC 呢?可以通过设置 Local-preference 属性来改变 RTD 选路的结果。

```
[RTB-bgp]default local-preference 0
[RTD]dis ip rou
Routing Tables: Public
         Destinations : 21        Routes : 22
Destination/Mask    Proto    Pre   Cost      NextHop        Interface
1.1.1.1/32          BGP      255   0         11.4.1.1       S2/0
2.2.2.2/32          OSPF     10    1562      11.3.1.1       S1/0
3.3.3.3/32          OSPF     10    1562      11.4.1.1       S2/0
4.4.4.4/32          Direct   0     0         127.0.0.1      InLoop0
10.0.0.0/8          BGP      255   0         11.4.1.1       S2/0
10.4.2.0/24         Direct   0     0         10.4.2.1       Eth0/1
10.4.2.1/32         Direct   0     0         127.0.0.1      InLoop0
11.1.0.0/16         OSPF     10    3124      11.3.1.1       S1/0
11.1.1.2/32         BGP      255   0         11.4.1.1       S2/0
11.2.0.0/16         OSPF     10    3124      11.4.1.1       S2/0
11.2.1.2/32         BGP      255   0         11.3.1.1       S1/0
11.3.0.0/16         Direct   0     0         11.3.1.2       S1/0
11.3.1.1/32         Direct   0     0         11.3.1.1       S1/0
11.3.1.2/32         Direct   0     0         127.0.0.1      InLoop0
11.4.0.0/16         BGP      255   0         11.3.1.1       S1/0
11.4.1.0/24         Direct   0     0         11.4.1.2       S2/0
11.4.1.1/32         Direct   0     0         11.4.1.1       S2/0
11.4.1.2/32         Direct   0     0         127.0.0.1      InLoop0
11.5.0.0/16         OSPF     10    1563      11.3.1.1       S1/0
                    OSPF     10    1563      11.4.1.1       S2/0
127.0.0.0/8         Direct   0     0         127.0.0.1      InLoop0
127.0.0.1/32        Direct   0     0         127.0.0.1      InLoop0
```

查看 RTD 的路由表,看到 RTD 的下一跳地址改为 10.4.1.1。

```
[RTD]tracert 10.1.1.1
  traceroute to 10.1.1.1(10.1.1.1) 30 hops max,40 bytes packet, press CTRL_C to b
reak
 1   11.4.1.1 17 ms 17 ms 17 ms
 2   * *
[RTA]dis ip rou
Routing Tables: Public
         Destinations : 15        Routes : 15
Destination/Mask    Proto    Pre   Cost      NextHop        Interface
1.1.1.1/32          Direct   0     0         127.0.0.1      InLoop0
2.2.2.2/32          BGP      255   1         11.2.1.2       S1/0
```

3.3.3.3/32	BGP	255	1	11.1.1.2	S2/0
10.0.0.0/8	Direct	0	0	10.1.1.1	Eth0/1
10.1.1.1/32	Direct	0	0	127.0.0.1	InLoop0
11.1.0.0/16	Direct	0	0	11.1.1.1	S2/0
11.1.1.1/32	Direct	0	0	127.0.0.1	InLoop0
11.1.1.2/32	Direct	0	0	11.1.1.2	S2/0
11.2.0.0/16	Direct	0	0	11.2.1.1	S1/0
11.2.1.1/32	Direct	0	0	127.0.0.1	InLoop0
11.2.1.2/32	Direct	0	0	11.2.1.2	S1/0
11.3.0.0/16	BGP	255	1563	11.2.1.2	S1/0
11.4.0.0/16	BGP	255	0	11.1.1.2	S2/0
127.0.0.0/8	Direct	0	0	127.0.0.1	InLoop0
127.0.0.1/32	Direct	0	0	127.0.0.1	InLoop0

从以上信息可以看出，自 RTA 去往 RTD 的 AS200 希望所有来自 RTA 去往 RTD 的路由为 RTB，要想使 RTA 去往 RTD 的 AS200 希望所有来自 RTA 去往 RTD 的路由为优选 RTC，可以通过 RTB 和 RTC 通告路由时附加不同的 MED 值实现。

```
[RTB-bgp]default med 150
[RTC-bgp]default med 100
[RTA]dis ip rou
Routing Tables: Public
        Destinations : 15      Routes : 15
```

Destination/Mask	Proto	Pre	Cost	NextHop	Interface
1.1.1.1/32	Direct	0	0	127.0.0.1	InLoop0
2.2.2.2/32	BGP	255	100	11.2.1.2	S1/0
3.3.3.3/32	BGP	255	150	11.1.1.2	S2/0
10.0.0.0/8	Direct	0	0	10.1.1.1	Eth0/1
10.1.1.1/32	Direct	0	0	127.0.0.1	InLoop0
11.1.0.0/16	Direct	0	0	11.1.1.1	S2/0
11.1.1.1/32	Direct	0	0	127.0.0.1	InLoop0
11.1.1.2/32	Direct	0	0	11.1.1.2	S2/0
11.2.0.0/16	Direct	0	0	11.2.1.1	S1/0
11.2.1.1/32	Direct	0	0	127.0.0.1	InLoop0
11.2.1.2/32	Direct	0	0	11.2.1.2	S1/0
11.3.0.0/16	BGP	255	100	11.2.1.2	S1/0
11.4.0.0/16	BGP	255	100	11.2.1.2	S1/0
127.0.0.0/8	Direct	0	0	127.0.0.1	InLoop0
127.0.0.1/32	Direct	0	0	127.0.0.1	InLoop0

RTA 的路由表发生了变化，去往 11.4.0.0 网段的下一跳地址由 11.1.1.2 改为 11.2.1.2，，也就是实现了所有来自 RTA 去往 RTD 的路由为优选 RTC，为什么会改变呢？让我们查看一下 RTA 的 BGP 路由表。

```
[RTA]dis bgp routing7:36:52 2009 RTB SHE
  Total Number of Routes: 16m con0
  BGP Local router ID is 1.1.1.1
  Status codes: * - valid, > - best, d - damped,
```

h - history,　i - internal, s - suppressed, S - Stale
Origin : i - IGP, e - EGP, ? - incomplete

	Network	NextHop	MED	LocPrf	PrefVal	Path/Ogn
*>	1.1.1.1/32	0.0.0.0	0		0	?
*>	2.2.2.2/32	11.2.1.2	100		0	200?
*>	3.3.3.3/32	11.1.1.2	150		0	200?
*>	10.0.0.0	0.0.0.0	0		0	i
*>	10.1.1.1/32	0.0.0.0	0		0	?
*>	11.1.0.0/16	0.0.0.0	0		0	?
*		11.2.1.2	100		0	200?
*>	11.1.1.1/32	0.0.0.0	0		0	?
*>	11.1.1.2/32	0.0.0.0	0		0	?
*>	11.2.0.0/16	0.0.0.0	0		0	?
*		11.1.1.2	150		0	200?
*>	11.2.1.1/32	0.0.0.0	0		0	?
*>	11.2.1.2/32	0.0.0.0	0		0	?
*>	11.3.0.0/16	11.2.1.2	100		0	200?
*>	11.4.0.0/16	11.2.1.2	100		0	200?
*		11.1.1.2	150		0	200?

去往 11.4.0.0/16 有两条路由，经过 RTC（11.2.1.2）路由的 MED 属性值为 100，经过 RTB 路由的 MED 属性值为 150。根据 BGP 的选路原则，RTA 优选 MED 值小的路由。

19.4　项目总结与提高

写出主要项目实施规划、步骤与实训所得的主要结论。

项目二十 BGP 路由反射

20.1 项目提出

小王在某公司上班,想了解 BGP 协议 LOCAL-PREF 与 MED 属性的应用,于是他设计了如下实验。

20.2 项目分析

1. 项目实训目的

理解路由反射器的原理、作用;
掌握路由反射器的配置。

2. 项目实现功能

通过实训,理解路由反射器的原理、作用,掌握路由反射器的配置。

3. 项目主要应用的技术介绍

在一个 BGP 域中,一个路由器通过 IBGP 从另一个路由器学习到的路由信息是不会转发给下一个 IBGP 路由器的,这是为了避免在 AS 中产生路由环路。则如果要想让下一个路由器学习到该路由信息,则产生该信息的源路由器必须与那个路由器再建立 IBGP 邻接关系。也就是说在同一个 BGP 域中,要想让所有路由器学习到所有的路由信息,则它们之间必须建立全网状的 IBGP 互联。显而易见,这样对于网络的扩展非常不利。为了克服这个问题,我们定义了路由反射器的概念。一台被配置为路由反射器的路由器一旦收到一条路由信息,它就会将这条路由信息传递给所有跟它建立客户关系的路由器,从而消除了对全互联环境的要求。

20.3 项目实施

1. 项目拓扑图

BGP 路由反射如图 20-1 所示。

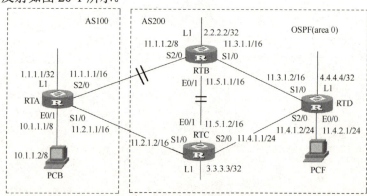

图 20-1 BGP 路由反射

2. 项目实训环境准备

MSR20-40（4台），计算机（2台）、网线若干条。CMW 版本：5.20。

为了不受原来的配置影响，在实训之前先将所有的配置数据擦除后重新启动，命令为：

<h3c>Reset saved-configuration
<h3c>reboot

3. 项目主要实训步骤

（1）此实训环境和配置与项目九相同，配置完成后，断开 RTB 和 RTC 之间的物理连接，断开 RTA 和 RTB 之间的连接。

[RTB]dis bgp peer
　BGP local router ID : 2.2.2.2
　Local AS number : 200
　Total number of peers : 3　　　　　　Peers in established state : 1
　Peer　　　　　　　AS　　MsgRcvd　MsgSent OutQ PrefRcv Up/Down　　State
　11.1.1.1　　　　100　　0　　　　0　　　　0　　0　　00:05:24　Active
　11.3.1.2　　　　200　　75　　　121　　　0　　3　　01:11:15　Established
　11.5.1.2　　　　200　　0　　　　0　　　　0　　0　　00:03:12　Active

只有一对连接开始"Established"状态。

[RTB]ping 10.1.1.1
　　PING 10.1.1.1: 56　data bytes, press CTRL_C to break
　　　Request time out
　　　Request time out
　　　Request time out
　　　Request time out
　　　Request time out
　　--- 10.1.1.1 ping statistics ---
　　　5 packet(s) transmitted
　　　0 packet(s) received
　100.00% packet loss

RTB 不能 ping 通 RTA。

[RTB]dis bgp routing
　Total Number of Routes: 7
　BGP Local router ID is 2.2.2.2
　Status codes: * - valid, > - best, d - damped,
　　　　　　　　h - history,　i - internal, s - suppressed, S - Stale
　　　　　　　　Origin : i - IGP, e - EGP, ? - incomplete

Network	NextHop	MED	LocPrf	PrefVal	Path/Ogn
* i 2.2.2.2/32	11.3.1.2	1562	100	0	?
*> 3.3.3.3/32	0.0.0.0	150	0	0	?
* i	11.3.1.2	1562	100	0	?
*> 11.2.0.0/16	0.0.0.0	150	0	0	?
* i	11.3.1.2	3124	100	0	?
*> 11.4.0.0/16	127.0.0.1			0	?

| | s> | 11.4.1.0/24 | 0.0.0.0 | 150 | 0 | 0 | ? |

没有去 10.0.0.0 网段的路由，所以不能 ping 通。根据 BGP 传播的原理：从 IBGP 获得的路由不向 IBGP 传播，RTD 不会通告给 RTB 10.0.0.0 网段的这条路由。这里我们采用另一种方式避免建立 BGP 连接消耗资源——BGP 反射。

（2）配置反射器。

```
[RTD-bgp]peer as200 reflect-client
[RTB]ping 10.1.1.1
  PING 10.1.1.1: 56    data bytes, press CTRL_C to break
    Reply from 10.1.1.1: bytes=56 Sequence=1 ttl=253 time=77 ms
    Reply from 10.1.1.1: bytes=56 Sequence=2 ttl=253 time=76 ms
    Reply from 10.1.1.1: bytes=56 Sequence=3 ttl=253 time=77 ms
    Reply from 10.1.1.1: bytes=56 Sequence=4 ttl=253 time=77 ms
    Reply from 10.1.1.1: bytes=56 Sequence=5 ttl=253 time=77 ms
  --- 10.1.1.1 ping statistics ---
    5 packet(s) transmitted
    5 packet(s) received
    0.00% packet loss
round-trip min/avg/max = 76/76/77 ms
[RTB]dis bgp routing
 Total Number of Routes: 9
 BGP Local router ID is 2.2.2.2
 Status codes: * - valid, > - best, d - damped,
               h - history,  i - internal, s - suppressed, S - Stale
               Origin : i - IGP, e - EGP, ? - incomplete
```

	Network	NextHop	MED	LocPrf	PrefVal	Path/Ogn
*>i	1.1.1.1/32	11.4.1.1	0	100	0	100?
* i	2.2.2.2/32	11.3.1.2	1562	100	0	?
*>	3.3.3.3/32	0.0.0.0	150	0	0	?
* i		11.3.1.2	1562	100	0	?
*>i	10.0.0.0	11.4.1.1	0	100	0	100i
*>	11.2.0.0/16	0.0.0.0	150	0	0	?
* i		11.3.1.2	3124	100	0	?
*>	11.4.0.0/16	127.0.0.1			0	?
s>	11.4.1.0/24	0.0.0.0	150	0	0	?

路由表中有了去 10.0.0.0 网段的路由，下一跳地址为 11.4.1.1（RTC）。

20.4 项目总结与提高

写出主要项目实施规划、步骤与实训所得的主要结论。

项目二十一 基于 AS_PATH 的路由策略

21.1 项目提出

小王在某公司上班,想了解基于 AS_PATH 的路由策略,于是他设计了如下实验。

21.2 项目分析

1. 项目实训目的

掌握 as_path 实现路由策略。

2. 项目实现功能

通过实训,掌握 as_path 实现路由策略。

3. 项目主要应用的技术介绍

路由策略是为了改变网络流量所经过的途径而修改路由信息的技术,主要通过改变路由属性(包括可达性)来实现。

在 BGP 的路由策略实施中,as-path 是最常使用的过滤方法之一。

as-path 是 BGP 协议中的一个重要的路径属性(Path attribute),as-path 顺序记录了一条 BGP 路由从源 AS 到目的 AS 所经过的路径(由源 AS 到目的 AS 所途径的所有 as-number 组成了一个字符串,其中不包含目的 as-number 自治系统号),每一条 BGP 路由都携带 as-path 属性,所以可以通过对 as-path 属性过滤来实现对 BGP 路由的过滤。

21.3 项目实施

1. 项目拓扑图

基于 AS_PATH 的路由策略如图 21-1 所示。

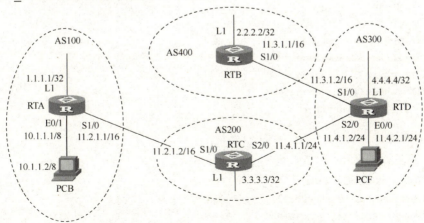

图 21-1 基于 AS_PATH 的路由策略

2. 项目实训环境准备

MSR20-40（4台），计算机（2台）、网线若干条。CMW 版本：5.20。

为了不受原来的配置影响，在实训之前先将所有的配置数据擦除后重新启动，命令为：

    \<h3c\>Reset saved-configuration
    \<h3c\>reboot

3. 项目主要实训步骤

【实训要求】

RTC 不向 RTA 通告 AS300 内的任何路由。

【实训步骤】

（1）完成实训的基本配置。

①RTA 的配置。

    [RTA]bgp 100
    [RTA-bgp]group as200 external
    [RTA-bgp]peer as200 as-number 200
    [RTA-bgp]peer 11.2.1.2 group as200
    [RTA-bgp]import-route direct
    [RTA-bgp]

②RTC 的配置。

    [RTB]bgp 200
    [RTC-bgp]group as100 external
    [RTC-bgp]group as300 external
    [RTC-bgp]peer as100 as-number 100
    [RTC-bgp]peer as300 as-number 300
    [RTC-bgp]peer 11.2.1.1 group as100
    [RTC-bgp]peer 11.4.1.2 group as400
    [RTC-bgp]import-route direct

③RTD 的配置。

    [RTD]bgp 300
    [RTD-bgp]group as200 external
    [RTD-bgp]peer as200 as-number 200
    [RTD-bgp]peer as400 as-number 400
    [RTD-bgp]peer 11.4.1.1 group as200
    [RTD-bgp]peer 11.3.1.1 group as400
    [RTD-bgp]import-route direct

④RTB 的配置。

    [RTB]bgp 400
    [RTB-bgp]group as300 external
    [RTB-bgp]peer as300 as-number 300
    [RTB-bgp]peer 11.3.1.2 group as300
    [RTB-bgp]import-route direct

项目二十一 基于 AS_PATH 的路由策略

在 RTB 上 ping 10.1.1.1,信息如下:

[RTB]ping 10.1.1.1
  PING 10.1.1.1: 56 data bytes, press CTRL_C to break
    Reply from 10.1.1.1: bytes=56 Sequence=1 ttl=253 time=77 ms
    Reply from 10.1.1.1: bytes=56 Sequence=2 ttl=253 time=77 ms
    Reply from 10.1.1.1: bytes=56 Sequence=3 ttl=253 time=77 ms
    Reply from 10.1.1.1: bytes=56 Sequence=4 ttl=253 time=77 ms
    Reply from 10.1.1.1: bytes=56 Sequence=5 ttl=253 time=77 ms
  --- 10.1.1.1 ping statistics ---
    5 packet(s) transmitted
    5 packet(s) received
    0.00% packet loss
  round-trip min/avg/max = 77/77/77 ms

(2) 配置 RTC 的路由策略。

在 RTC 上配置路由策略,使其不向 RTA 通告 AS300 内的任何路由,但不包括 AS 400 内的路由。

①查看配置路由策略前的 RTC 的 BGP 路由表。

[RTC]dis bgp routing
 Total Number of Routes: 18
 BGP Local router ID is 3.3.3.3
 Status codes: * - valid, > - best, d - damped,
               h - history, i - internal, s - suppressed, S - Stale
 Origin : i - IGP, e - EGP, ? - incomplete

	Network	NextHop	MED	LocPrf	PrefVal	Path/Ogn
*>	1.1.1.1/32	11.2.1.1	0		0	100?
*>	2.2.2.2/32	11.4.1.2			0	300 400?
*>	3.3.3.3/32	0.0.0.0	0		0	?
*>	4.4.4.4/32	11.4.1.2	0		0	300?
*>	10.0.0.0	11.2.1.1	0		0	100?
*>	10.4.2.0/24	11.4.1.2	0		0	300?
*>	11.2.0.0/16	0.0.0.0	0		0	?
*		11.2.1.1	0		0	100?
*>	11.2.1.1/32	0.0.0.0	0		0	?
*>	11.2.1.2/32	0.0.0.0	0		0	?
*		11.2.1.1	0		0	100?
*>	11.3.0.0/16	11.4.1.2	0		0	300?
*>	11.3.1.1/32	11.4.1.2	0		0	300?
*>	11.4.1.0/24	0.0.0.0	0		0	?
*		11.4.1.2	0		0	300?
*>	11.4.1.1/32	0.0.0.0	0		0	?
*		11.4.1.2	0		0	300?
*>	11.4.1.2/32	0.0.0.0	0		0	?

②在 RTC 上配置策略路由:

[RTC]ip as-path 1 deny ^300$

[RTC]ip as-path 1 permit .*
[RTC-bgp]peer as100 as-path 1 export

③配置好策略路由查看 RTA 的 BGP 路由表：

[RTA]display BGP routing
Total Number of Routes: 12
BGP Local router ID is 1.1.1.1
Status codes: * - valid, > - best, d - damped,
　　　　　　 h - history,　i - internal, s - suppressed, S - Stale
Origin : i - IGP, e - EGP, ? - incomplete

	Network	NextHop	MED	LocPrf	PrefVal	Path/Ogn
*>	1.1.1.1/32	0.0.0.0	0		0	?
*>	2.2.2.2/32	11.2.1.2			0	200 300 400?
*>	3.3.3.3/32	11.2.1.2	0		0	200?
*>	10.0.0.0	0.0.0.0	0		0	?
*>	10.1.1.1/32	0.0.0.0	0		0	?
*>	11.2.0.0/16	0.0.0.0	0		0	?
*		11.2.1.2	0		0	200?
*>	11.2.1.1/32	0.0.0.0	0		0	?
*		11.2.1.2	0		0	200?
*>	11.2.1.2/32	0.0.0.0	0		0	?
*>	11.4.1.0/24	11.2.1.2	0		0	200?
*>	11.4.1.2/32	11.2.1.2	0		0	200?

你会发现始发于 AS400 的路由被 RTC 通告给 RTA 了，但始发于 AS300 的路由没有被通告。

④查看 RTD 的 BGP 路由表。

[RTD]dis bgp routing
Total Number of Routes: 19
BGP Local router ID is 4.4.4.4
Status codes: * - valid, > - best, d - damped,
　　　　　　 h - history,　i - internal, s - suppressed, S - Stale
Origin : i - IGP, e - EGP, ? - incomplete

	Network	NextHop	MED	LocPrf	PrefVal	Path/Ogn
*>	1.1.1.1/32	11.4.1.1			0	200 100?
*>	2.2.2.2/32	11.3.1.1	0		0	400?
*>	3.3.3.3/32	11.4.1.1	0		0	200?
*>	4.4.4.4/32	0.0.0.0	0		0	?
*>	10.0.0.0	11.4.1.1			0	200 100?
*>	10.4.2.0/24	0.0.0.0	0		0	?
*>	10.4.2.1/32	0.0.0.0	0		0	?
*>	11.2.0.0/16	11.4.1.1	0		0	200?
*>	11.2.1.1/32	11.4.1.1	0		0	200?
*>	11.3.0.0/16	0.0.0.0	0		0	?
*		11.3.1.1	0		0	400?
*>	11.3.1.1/32	0.0.0.0	0		0	?

*>	11.3.1.2/32	0.0.0.0	0	0	?
*		11.3.1.1	0	0	400?
*>	11.4.1.0/24	0.0.0.0	0	0	?
*		11.4.1.1	0	0	200?
*>	11.4.1.1/32	0.0.0.0	0	0	?
*>	11.4.1.2/32	0.0.0.0	0	0	?
*		11.4.1.1	0	0	200?

始发于 AS100 的路由被正常通告给 RTD。

⑤从 RTB 的 loopback1 2.2.2.2 ping 10.1.1.1。

  [RTB]ping -a 2.2.2.2 10.1.1.1
   PING 10.1.1.1: 56 data bytes, press CTRL_C to break
    Reply from 10.1.1.1: bytes=56 Sequence=1 ttl=253 time=78 ms
    Reply from 10.1.1.1: bytes=56 Sequence=2 ttl=253 time=76 ms
    Reply from 10.1.1.1: bytes=56 Sequence=3 ttl=253 time=78 ms
    Reply from 10.1.1.1: bytes=56 Sequence=4 ttl=253 time=76 ms
    Reply from 10.1.1.1: bytes=56 Sequence=5 ttl=253 time=77 ms
   --- 10.1.1.1 ping statistics ---
   5 packet(s) transmitted
   5 packet(s) received
   0.00% packet loss
   round-trip min/avg/max = 76/77/78 ms

能够 ping 通，由于 RTC 没有过滤始发于 AS100、AS400 的路由，所以 RTB 可以 ping 通 RTA。

⑥从 RTD 的 loopback1 4.4.4.4 ping 10.1.1.1。

  [RTD]ping -a 4.4.4.4 10.1.1.1
   PING 10.1.1.1: 56 data bytes, press CTRL_C to break
    Request time out
    Request time out
    Request time out
    Request time out
   --- 10.1.1.1 ping statistics ---
   4 packet(s) transmitted
   0 packet(s) received
   100.00% packet loss

RTD 无法 ping 通 RTA。

"^300$" 表示从匹配以 300 开始并以 300 结束的路径，因此将过滤 AS300 始发的路由信息。如将 "^300$" 改为 "^300"，则 acl 将拒绝 RTC 向 RTA 通告任何从 AS300 和从 AS400 始发的路由。

  [RTA]dis bgp rou
   Total Number of Routes: 12
   BGP Local router ID is 1.1.1.1
   Status codes: * - valid, > - best, d - damped,
        h - history, i - internal, s - suppressed, S - Stale

Origin : i - IGP, e - EGP, ? - incomplete

	Network	NextHop	MED	LocPrf	PrefVal	Path/Ogn
*>	1.1.1.1/32	0.0.0.0	0		0	?
*>	2.2.2.2/32	11.2.1.2			0	200 300 400?
*>	3.3.3.3/32	11.2.1.2	0		0	200?
*>	10.0.0.0	0.0.0.0	0		0	?
*>	10.1.1.1/32	0.0.0.0	0		0	?
*>	11.2.0.0/16	0.0.0.0	0		0	?
*		11.2.1.2	0		0	200?
*>	11.2.1.1/32	0.0.0.0	0		0	?
*		11.2.1.2	0		0	200?
*>	11.2.1.2/32	0.0.0.0	0		0	?
*>	11.4.1.0/24	11.2.1.2	0		0	200?
*>	11.4.1.2/32	11.2.1.2	0		0	200?

没有通往 RTC 的路由信息。

21.4 项目总结与提高

写出主要项目实施规划、步骤与实训所得的主要结论。

项目二十二　基于 Community 属性的路由策略

22.1　项目提出

小王在某公司上班，想了解基于 Community 属性的路由策略，于是他设计了如下实验。

22.2　项目分析

1. 项目实训目的

掌握 Community 属性实现路由策略。

2. 项目实现功能

通过实训，掌握 Community 属性实现路由策略。

3. 项目主要应用的技术介绍

团体属性列表（community-list），community-list 仅用于 BGP。BGP 的路由信息包中包含一个 community 属性域，用来标识一个团体。community-list 就是针对团体属性域指定匹配条件。

22.3　项目实施

1. 项目拓扑图

基于 Community 属性的路由策略如图 22-1 所示。

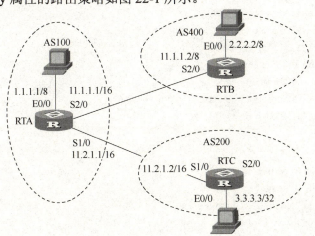

图 22-1　基于 Community 属性的路由策略

2. 项目实训环境准备

MSR20-40（3 台），计算机（2 台）、网线若干条。CMW 版本：5.20。

为了不受原来的配置影响,在实训之前先将所有的配置数据擦除后重新启动,命令为:

    <h3c>Reset saved-configuration
    <h3c>reboot

3. 项目主要实训步骤

【实训要求】

RTA、RTB 和 RTC 互为 BGP 对等体,其中 RTB 和 RTC 向对等体发布带有 Community 属性的路由信息,RTA 根据 community 属性设置该路由不向对等体发布。

【实训步骤】

(1)完成实训的基本配置(包括 IP 地址配置,设备命名)。
(2)BGP 和策略路由配置:
①RTA 的配置。

    [RTA]ip community-list 10 permit 100:200
    [RTA]ip community-list 10 permit 100:400
    [RTA]route-policy noexp permit node 10
    [RTA]route-policy noexp permit node 10
    Info: New Sequence of this List
    [RTA-route-policy]if-match community 10
    [RTA-route-policy]apply community no-export
    [RTA-route-policy]quit
    [RTA]route-policy noexp permit node 20
    Info: New Sequence of this List
    [RTA-route-policy]if-match community 20
    [RTA-route-policy]apply community no-export
    [RTA]bgp 100
    [RTA-bgp]network 1.0.0.0
    [RTA-bgp]group as200 external
    [RTA-bgp]peer as200 as-number 200
    [RTA-bgp]peer as200 route-policy noexp import
    [RTA-bgp]peer 11.2.1.2 group as200
    [RTA-bgp]group as400 external
    [RTA-bgp]peer as400 as-number 400
    [RTA-bgp]peer as400 route-policy noexp import
    [RTA-bgp]peer 11.1.1.2 group as400

②RTB 的配置。

    [RTB]ip as-path 10 permit ^$
    [RTB]ip as-path 10 deny .*
    [RTB]route-policy community permit node 10
    Info: New Sequence of this List
    [RTB-route-policy]if-match as-path 10
    [RTB-route-policy]apply community 100:400
    [RTB-route-policy]quit
    [RTB]bgp 400
    [RTB-bgp]network 2.0.0.0

项目二十二　基于 Community 属性的路由策略

　　[RTB-bgp]group as100 external
　　[RTB-bgp]peer as100 as-number 100
　　[RTB-bgp]peer as100 route-policy community export
　　[RTB-bgp]peer as100 advertise-community
　　[RTB-bgp]peer 11.1.1.1 group as100

③RTC 的配置。

　　[RTC]ip as-path 10 permit ^$
　　[RTC]ip as-path 10 deny .*
　　[RTC]route-policy community permit node 10
　　Info: New Sequence of this List
　　[RTC-route-policy]if-match as-path 10
　　[RTC-route-policy]apply community 100:200
　　[RTC-route-policy]quit
　　[RTC]bgp 200
　　[RTC-bgp]network 3.0.0.0
　　[RTC-bgp]group as100 external
　　[RTC-bgp]peer as100 as-number 100
　　[RTC-bgp]peer as100 route-policy community export
　　[RTC-bgp]peer as100 advertise-community
　　[RTC-bgp]peer 11.2.1.1 group as100

④此时，对等体关系已经建立：

　　[RTA]dis bgp peer
　　BGP local router ID : 1.1.1.1
　　Local AS number : 100
　　Total number of peers : 2　　　　　　Peers in established state : 2
　　Peer　　　　　　　　AS　MsgRcvd　MsgSent OutQ PrefRcv Up/Down　　State
　　11.2.1.2　　　　　　200　　4　　　　4　　0　　1 00:01:11 Established
　　11.1.1.2　　　　　　400　　9　　　　9　　0　　1 00:07:03 Established

⑤查看 RTA、RTB 和 RTC 的路由表。

　　[RTA]dis bgp routing
　　Total Number of Routes: 3
　　BGP Local router ID is 1.1.1.1
　　Status codes: * - valid, > - best, d - damped,
　　　　　　　　　h - history,　i - internal, s - suppressed, S - Stale
　　　　　　　　　Origin : i - IGP, e - EGP, ? - incomplete
　　　　　Network　　　　NextHop　　　　MED　　　　LocPrf　　　PrefVal Path/Ogn
　　　*>　1.0.0.0　　　　0.0.0.0　　　　0　　　　　　　　　　　0　　　i
　　　*>　2.0.0.0　　　　11.1.1.2　　　　0　　　　　　　　　　　0　　　400i
　　　*>　3.0.0.0　　　　11.2.1.2　　　　0　　　　　　　　　　　0　　　200i
　　[RTB]dis bgp rou
　　Total Number of Routes: 2
　　BGP Local router ID is 11.1.1.2
　　Status codes: * - valid, > - best, d - damped,
　　　　　　　　　h - history,　i - internal, s - suppressed, S - Stale

Origin : i - IGP, e - EGP, ? - incomplete

	Network	NextHop	MED	LocPrf	PrefVal	Path/Ogn
*>	1.0.0.0	11.1.1.1	0		0	100i
*>	2.0.0.0	0.0.0.0	0		0	i

[RTC]dis bgp routing
Total Number of Routes: 2
BGP Local router ID is 11.2.1.2
Status codes: * - valid, > - best, d - damped,
　　　　　　 h - history, i - internal, s - suppressed, S - Stale
Origin : i - IGP, e - EGP, ? - incomplete

	Network	NextHop	MED	LocPrf	PrefVal	Path/Ogn
*>	1.0.0.0	11.2.1.1	0		0	100i
*>	3.0.0.0	0.0.0.0	0		0	i

　　RTB/RTC 的配置为将向外发布的所有路由的 Community 属性分别配置为 100：400 和 100：200；RTA 的配置为将从 AS200 及 AS400 收到得路由的 Community 类型配置为 no-export，即不再向 EBGP 对等体发布此路由信息。所以配置的结果是 RTA 上有到 2.0.0.0/8 和 3.0.0.0/8 的路由，但 RTB 上没有到 3.0.0.0/8 的路由，RTC 上也没有到 2.0.0.0/8 的路由。

22.4　项目总结与提高

　　写出主要项目实施规划、步骤与实训所得的主要结论。

项目二十三　引入其他路由协议

23.1　项目提出

路由器连接了一所大学的校园网和一个地区性网络。校园网使用 RIP 作为其内部路由协议，地区性网络使用 OSPF 路由协议，路由器需要将校园网中的某些路由信息在地区性网络中发布。为实现这一功能，路由器上的 OSPF 协议在引入 RIP 协议路由信息时通过对一个路由策略的引入实现路由过滤功能。该路由器由两个节点组成，实现 192.1.0.0/24 和 128.2.0.0/16 的路由信息以不同的路由权值被 OSPF 协议发布。

23.2　项目分析

1．项目实训目的

掌握路由策略的基本配置；
掌握引入路由的其他方法。

2．项目实现功能

通过实训，掌握路由策略的基本配置，掌握引入路由的其他方法。

3．项目主要应用的技术介绍

路由引入的原因：
（1）路由信息共享。
（2）不同 AS 运行不同路由协议的路由器之间为了获得对方的路由信息，必须在起边界网关作用的路由上引入路由。
（3）路由引入使支持不同路由协议的路由器在网络中协同工作成为可能。

23.3　项目实施

1．项目拓扑图

引入其他路由协议如图 23-1 所示。

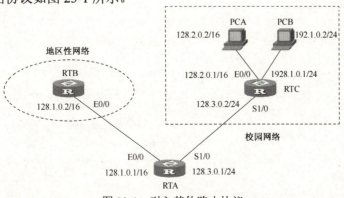

图 23-1　引入其他路由协议

2. 项目实训环境准备

MSR20-40（3台），计算机（2台）、网线若干条。CMW 版本：5.20。

为了不受原来的配置影响，在实训之前先将所有的配置数据擦除后重新启动，命令为：

<h3c>Reset saved-configuration
<h3c>reboot

3. 项目主要实训步骤

（1）按拓扑图连接设备，并完成实训的基本配置（包括 IP 地址配置，设备命名）。

①RTA 的配置。

```
<H3C>sys
[H3C]sysn RTA
[RTA]dis ip int b
*down: administratively down
(s): spoofing
```

Interface	Physical	Protocol	IP Address	Description
Aux0	down	down	unassigned	Aux0 Inte...
Ethernet0/0	up	down	unassigned	Ethernet0...
Ethernet0/1	down	down	unassigned	Ethernet0...
Serial1/0	up	up	unassigned	Serial1/0...
Serial2/0	down	down	unassigned	Serial2/0...

```
[RTA]int s1/0
[RTA]int s1/0
[RTA-Serial1/0]ip address 128.3.0.1 24
[RTA-Serial1/0]int e0/0
[RTA-Ethernet0/0]ip address 128.1.0.1 16
[RTA-Ethernet0/0]
%Aug   8 08:10:10:863 2010 RTA IFNET/4/UPDOWN:
    Line protocol on the interface Ethernet0/0 is UP
```

②RTB 的配置。

```
<H3C>sys
<H3C>system-view
System View: return to User View with Ctrl+Z.
[H3C]sysname RTB
[RTB]dis ip int b
*down: administratively down
(s): spoofing
```

Interface	Physical	Protocol	IP Address	Description
Aux0	down	down	unassigned	Aux0 Inte...
Ethernet0/0	up	up	unassigned	Ethernet0...
Ethernet0/1	down	down	unassigned	Ethernet0...
Serial1/0	down	down	unassigned	Serial1/0...
Serial2/0	down	down	unassigned	Serial2/0...

```
[RTB]int e0/0
[RTB-Ethernet0/0]ip address 128.1.0.2 16
```

[RTB-Ethernet0/0]dis ip int b
*down: administratively down
(s): spoofing

Interface	Physical	Protocol	IP Address	Description
Aux0	down	down	unassigned	Aux0 Inte...
Ethernet0/0	up	up	128.1.0.2	Ethernet0...
Ethernet0/1	down	down	unassigned	Ethernet0...
Serial1/0	down	down	unassigned	Serial1/0...
Serial2/0	down	down	unassigned	Serial2/0...

[RTB-Ethernet0/0]

③RTC 的配置。

<H3C>system-view
System View: return to User View with Ctrl+Z.
[H3C]sysname RTC
[RTC]dis ip int b
*down: administratively down
(s): spoofing

Interface	Physical	Protocol	IP Address	Description
Aux0	down	down	unassigned	Aux0 Inte...
Ethernet0/0	up	up	unassigned	Ethernet0...
Ethernet0/1	up	down	unassigned	Ethernet0...
Serial1/0	up	up	unassigned	Serial1/0...
Serial2/0	down	down	unassigned	Serial2/0...

[RTC]int s1/0
[RTC-Serial1/0]ip address 128.3.0.2 24
[RTC-Serial1/0]
%Aug 8 08:11:14:882 2010 RTC IFNET/4/UPDOWN:
    Protocol PPP IPCP on the interface Serial1/0 is UP
[RTC-Serial1/0]int e0/0
[RTC-Ethernet0/0]ip address 128.2.0.1 16
[RTC-Ethernet0/0]int e0/1
[RTC-Ethernet0/1]ip address 192.1.0.1 24
[RTC-Ethernet0/1]
%Aug 8 08:12:39:394 2010 RTC IFNET/4/UPDOWN:
    Line protocol on the interface Ethernet0/1 is UP
[RTC-Ethernet0/1]dis ip int b
*down: administratively down
(s): spoofing

Interface	Physical	Protocol	IP Address	Description
Aux0	down	down	unassigned	Aux0 Inte...
Ethernet0/0	up	up	128.2.0.1	Ethernet0...
Ethernet0/1	up	up	192.1.0.1	Ethernet0...
Serial1/0	up	up	128.3.0.2	Serial1/0...
Serial2/0	down	down	unassigned	Serial2/0...

[RTC-Ethernet0/1]

(2) 对各路由器分别进行配置如下：
①RTA 的配置。

    [RTA]ospf 1
    [RTA-ospf-1]area 0.0.0.0
    [RTA-ospf-1-area-0.0.0.0]network 128.1.0.1 0.0.255.255
    [RTA-ospf-1-area-0.0.0.0]quit
    [RTA-ospf-1]quit
    [RTA]rip
    [RTA-rip-1]undo summary
    [RTA-rip-1]network 128.3.0.1
    [RTA-rip-1]quit
    [RTA]int s1/0
    [RTA-Serial1/0]rip version 2
    [RTA-Serial1/0]quit
    [RTA]ip ip-prefix p1 index 10 permit 192.1.0.0 24
    [RTA]ip ip-prefix p2 index 10 permit 128.2.0.0 16
    [RTA]route-policy r1 permit node 10
    [RTA-route-policy]if-match ip-prefix p1
    [RTA-route-policy]apply cost 120
    [RTA-route-policy]quit
    [RTA]route-policy r1 permit node 20
    Info: New Sequence of this List
    [RTA-route-policy]if-match ip-prefix p2
    [RTA-route-policy]apply cost 100
    [RTA-route-policy]quit

②RTB 的配置。

    [RTB]ospf
    [RTB-ospf-1]area 0
    [RTB-ospf-1-area-0.0.0.0]network 128.1.0.2 0.0.255.255
    [RTB-ospf-1-area-0.0.0.0]quit
    [RTB-ospf-1]quit

③RTC 的配置。

    [RTC]rip
    [RTC-rip-1]undo summary
    [RTC-rip-1]network 128.3.0.1
    [RTC-rip-1]network 128.2.0.1
    [RTC-rip-1]network 192.1.0.1
    [RTC-rip-1]quit
    [RTC]int e0/0
    [RTC-Ethernet0/0]rip version 2
    [RTC-Ethernet0/0]int e0/1
    [RTC-Ethernet0/1]rip version 2
    [RTC-Ethernet0/1]int s1/0
    [RTC-Serial1/0]rip version 2

(3) 观察现象。

在 RTA 上面查看路由表信息:

[RTA]display ip routing
Routing Tables: Public
    Destinations : 9   Routes : 9

Destination/Mask	Proto	Pre	Cost	NextHop	Interface
127.0.0.0/8	Direct	0	0	127.0.0.1	InLoop0
127.0.0.1/32	Direct	0	0	127.0.0.1	InLoop0
128.1.0.0/16	Direct	0	0	128.1.0.1	Eth0/0
128.1.0.1/32	Direct	0	0	127.0.0.1	InLoop0
128.2.0.0/16	RIP	100	1	128.3.0.2	S1/0
128.3.0.0/24	Direct	0	0	128.3.0.1	S1/0
128.3.0.1/32	Direct	0	0	127.0.0.1	InLoop0
128.3.0.2/32	Direct	0	0	128.3.0.2	S1/0
192.1.0.0/24	RIP	100	1	128.3.0.2	S1/0

发现 RIP 路由信息已经学习完毕,并且其优先级为 100,这是 RIP 路由协议在路由器中默认的优先级别。

在 RTA 中同时启用了 OSPF 路由协议,并在 OSPF 中对 RIP 路由进行了引入操作。当 RIP 路由被 OSPF 协议引入后,RTA 会将引入的路由信息发布给对 OSPF 的邻居,即 RTB。

查看 RTB 的路由信息:

[RTA]ospf
[RTA-ospf-1]import-route rip route-policy r1
[RTB]dis ip rou
Routing Tables: Public
    Destinations : 6   Routes : 6

Destination/Mask	Proto	Pre	Cost	NextHop	Interface
127.0.0.0/8	Direct	0	0	127.0.0.1	InLoop0
127.0.0.1/32	Direct	0	0	127.0.0.1	InLoop0
128.1.0.0/16	Direct	0	0	128.1.0.2	Eth0/0
128.1.0.2/32	Direct	0	0	127.0.0.1	InLoop0
128.2.0.0/16	O_ASE	150	100	128.1.0.1	Eth0/0
192.1.0.0/24	O_ASE	150	120	128.1.0.1	Eth0/0

在 RTA 发布时,路由信息的属性已经被改变。其实,这一过程是在 OSPF 协议引入 RIP 路由时就已经发生了。当引入路由协议时,如果匹配了指定的路由策略,则就已经开始执行路由策略的动作了。

23.4 项目总结与提高

写出主要项目实施规划、步骤与实训所得的主要结论。

项目二十四　OSPF 路由协议过滤接收的路由信息

24.1　项目提出

参考拓扑图，通过在 RTB 上配置路由过滤的规则，使接收到的三条静态路由部分可见，部分被屏蔽掉——20.0.0.0 和 40.0.0.0 网段的路由可见，30.0.0.0 网段的路由则被屏蔽。

24.2　项目分析

1. 项目实训目的

掌握路由策略的基本配置；
掌握过滤接收路由信息的方法。

2. 项目实现功能

通过实训，掌握路由策略的基本配置，掌握过滤接收路由信息的方法。

3. 项目主要应用的技术介绍

路由过滤可以对路由通告施加严格的控制，在路由更新中抑制某些路由不被发送和接收。
路由过滤器作为基本构建单元被用于创建路由选择策略（Routing Policy）。
路由选择策略是控制网络中的数据包如何转发以及改变数据包缺省转发属性的一组规则。
外部的路由可以进入到路由表中，路由表中的路由也可以被通告出去，那么路由过滤器正是通过管制这些出入路由表的路由来工作的。

OSPF 路由过滤技术分为：

- 对接收的路由进行过滤
- 对发布的路由进行过滤
- 对引入的路由进行过滤

24.3　项目实施

1. 项目拓扑图

OSPF 路由协议过滤接收的路由信息如图 24-1 所示。

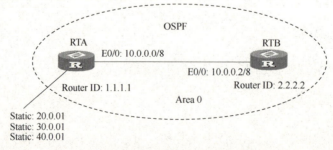

图 24-1　OSPF 路由协议过滤接收的路由信息

2. 项目实训环境准备

MSR20-40（2台），计算机（1台）、网线若干条。CMW 版本：5.20。

为了不受原来的配置影响，在实训之前先将所有的配置数据擦除后重新启动，命令为：

    \<h3c\>Reset saved-configuration
    \<h3c\>reboot

3. 项目主要实训步骤

（1）按拓扑图连接设备，并完成实训的基本配置（包括 IP 地址配置，设备命名）。

①RTA 的配置。

```
[H3C]sysname RTA
[RTA]dis ip int b
*down: administratively down
(s): spoofing
Interface              Physical Protocol IP Address    Description
Aux0                   down     down     unassigned    Aux0 Inte...
Ethernet0/0            up       down     unassigned    Ethernet0...
Ethernet0/1            down     down     unassigned    Ethernet0...
Serial1/0              down     down     unassigned    Serial1/0...
Serial2/0              down     down     unassigned    Serial2/0...
[RTA]int loopback1
[RTA-LoopBack1]ip address 1.1.1.1 32
[RTA-LoopBack1]quit
[RTA]int e0/0
[RTA-Ethernet0/0]ip add 10.0.0.1 8
[RTA-Ethernet0/0]
%Aug  8 09:45:26:489 2010 RTA IFNET/4/UPDOWN:
    Line protocol on the interface Ethernet0/0 is UP
```

②RTB 的配置。

```
[H3C]sysname RTB
[RTB]dis ip int b
*down: administratively down
(s): spoofing
Interface              Physical Protocol IP Address    Description
Aux0                   down     down     unassigned    Aux0 Inte...
Ethernet0/0            up       down     unassigned    Ethernet0...
Ethernet0/1            down     down     unassigned    Ethernet0...
Serial1/0              down     down     unassigned    Serial1/0...
Serial2/0              down     down     unassigned    Serial2/0...
[RTB]int e0/0
[RTB-Ethernet0/0]ip address 10.0.0.2 8
[RTB-Ethernet0/0]
%Feb 28 09:58:06:559 2009 RTB IFNET/4/UPDOWN:
    Line protocol on the interface Ethernet0/0 is UP
[RTB-Ethernet0/0]int loopback1
```

[RTB-LoopBack1]ip add 2.2.2.2 32
[RTB-LoopBack1] quit

（2）对各路由器分别进行配置如下：
①RTA 的配置。

[RTA]router id 1.1.1.1
[RTA]ospf 1
[RTA-ospf-1]area 0
[RTA-ospf-1-area-0.0.0.0]network 10.0.0.1 0.255.255.255
[RTA-ospf-1-area-0.0.0.0]quit
[RTA-ospf-1]import-route static
[RTA-ospf-1]quit
[RTA]ip route-static 20.0.0.0 24 null 0
[RTA]ip route-static 30.0.0.0 24 null 0
[RTA]ip route-static 40.0.0.0 24 null 0

②RTB 的配置。

[RTB]router id 2.2.2.2
[RTB]ospf
[RTB-ospf-1]area 0
[RTB-ospf-1-area-0.0.0.0]network 10.0.0.2 0.255.255.255
[RTB-ospf-1-area-0.0.0.0]quit
[RTB-ospf-1-area-0.0.0.0]quit
[RTB-ospf-1]quit
[RTB]acl number 2001
[RTB-acl-basic-2001]rule deny source 30.0.0.0 0.255.255.255
[RTB-acl-basic-2001]rule permit source any
[RTB-acl-basic-2001]quit
[RTB]

（3）观察现象。

配置完毕后，分别查看 RTA 和 RTB 的路由表：

[RTA]dis ip routing
Routing Tables: Public
        Destinations : 8   Routes : 8
Destination/Mask   Proto   Pre Cost   NextHop   Interface
1.1.1.1/32   Direct 0   0   127.0.0.1   InLoop0
10.0.0.0/8   Direct 0   0   10.0.0.1   Eth0/0
10.0.0.1/32   Direct 0   0   127.0.0.1   InLoop0
20.0.0.0/24   Static 60   0   0.0.0.0   NULL0
30.0.0.0/24   Static 60   0   0.0.0.0   NULL0
40.0.0.0/24   Static 60   0   0.0.0.0   NULL0
127.0.0.0/8   Direct 0   0   127.0.0.1   InLoop0
127.0.0.1/32   Direct 0   0   127.0.0.1   InLoop0
[RTB]dis ip routing-table
Routing Tables: Public
        Destinations : 8   Routes : 8

Destination/Mask	Proto	Pre	Cost	NextHop	Interface
2.2.2.2/32	Direct	0	0	127.0.0.1	InLoop0
10.0.0.0/8	Direct	0	0	10.0.0.2	Eth0/0
10.0.0.2/32	Direct	0	0	127.0.0.1	InLoop0
20.0.0.0/24	O_ASE	150	1	10.0.0.1	Eth0/0
30.0.0.0/24	O_ASE	150	1	10.0.0.1	Eth0/0
40.0.0.0/24	O_ASE	150	1	10.0.0.1	Eth0/0
127.0.0.0/8	Direct	0	0	127.0.0.1	InLoop0
127.0.0.1/32	Direct	0	0	127.0.0.1	InLoop0

此时，在 RTB 上面已经将静态路由映入到 OSPF 中。但是，我们已经配置了 ACL，但没有生效，要拒绝的网段 30.0.0.0 仍然被 OSPF 所接收了，这是为什么呢？因为我们仅仅配置了 ACL 的过滤规则，并没有将其应用起来，所以它并没有生效。在 RTB 上增加如下配置：

[RTB]ospf
[RTB-ospf-1]filter-policy 2001 import
[RTB-ospf-1]quit

再来查看 RTB 的路由表：

[RTB]dis ip routing-table
Routing Tables: Public
    Destinations : 7   Routes : 7

Destination/Mask	Proto	Pre	Cost	NextHop	Interface
2.2.2.2/32	Direct	0	0	127.0.0.1	InLoop0
10.0.0.0/8	Direct	0	0	10.0.0.2	Eth0/0
10.0.0.2/32	Direct	0	0	127.0.0.1	InLoop0
20.0.0.0/24	O_ASE	150	1	10.0.0.1	Eth0/0
40.0.0.0/24	O_ASE	150	1	10.0.0.1	Eth0/0
127.0.0.0/8	Direct	0	0	127.0.0.1	InLoop0
127.0.0.1/32	Direct	0	0	127.0.0.1	InLoop0

可以看出，30.0.0.0 网段已经被过滤掉了。我们再查看一下 OSPF 路由信息：

[RTB]dis ospf routing

    OSPF Process 1 with Router ID 2.2.2.2
      Routing Tables

Routing for Network
Destination	Cost	Type	NextHop	AdvRouter	Area
10.0.0.0/8	1	Transit	10.0.0.2	1.1.1.1	0.0.0.0

Routing for ASEs
Destination	Cost	Type	Tag	NextHop	AdvRouter
20.0.0.0/24	1	Type2	1	10.0.0.1	1.1.1.1
30.0.0.0/24	1	Type2	1	10.0.0.1	1.1.1.1
40.0.0.0/24	1	Type2	1	10.0.0.1	1.1.1.1

Total Nets: 4
 Intra Area: 1 Inter Area: 0 ASE: 3 NSSA: 0

发现这里面仍然包含 30.0.0.0 网段的路由信息，难道是产生了错误码？

我们知道，在 OSPF 学习过程中，20.0.0.0、30.0.0.0 和 40.0.0.0 网段的路由信息是都被发送的，而路由策略是针对路由表操作的，所以 OSPF 学习到 30.0.0.0 网段的路由信息后，才能对 30.0.0.0 网段进行过滤。

24.4 项目总结与提高

写出主要项目实施规划、步骤与实训所得的主要结论。

项目二十五 RIP 过滤发布路由信息

25.1 项目提出

路由器连接了校园网 A 和校园网 B，它们都使用 RIP 作为内部路由协议，路由器仅将校园网 A 中的 192.1.1.0/24 和 192.1.2.0/24 两个网段的路由发布到校园网 B 中去。为实现这一功能，路由器上的 RIP 协议使用一条 filter-policy 命令过滤发布的路由信息，通过对一个地址前缀列表的引用来对路由器发布的路由信息进行过滤。

25.2 项目分析

1. 项目实训目的

掌握路由策略的基本配置；
掌握过滤发布路由信息的方法。

2. 项目实现功能

通过实训，掌握路由策略的基本配置，掌握过滤发布路由信息的方法。

3. 项目主要应用的技术介绍

路由器在发布与接收路由信息时，可能需要对路由信息进行过滤。常用的路由过滤工具有 ACL、地址前缀列表等。

路由过滤方法：过滤路由协议报文；过滤路由协议报文中携带的路由信息；对 LSDB 计算出的路由信息进行过滤。

通过与 ACL、地址前缀列表结合使用，filter-policy 可对路由信息进行过滤。用 filter-policy 过滤 RIP 路由。

- 接收路由过滤：对进入 RIP 路由表的路由信息进行过滤。
- 发送路由过滤：对所发送的 RIP 路由信息进行过滤。

25.3 项目实施

1. 项目拓扑图

RIP 过滤发布路由信息如图 25-1 所示。

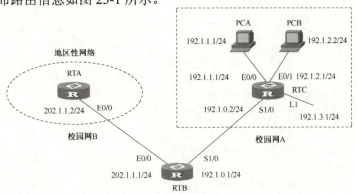

图 25-1 RIP 过滤发布路由信息

2. 项目实训环境准备

MSR20-40（3 台），计算机（2 台）、网线若干条。CMW 版本：5.20。

为了不受原来的配置影响，在实训之前先将所有的配置数据擦除后重新启动，命令为：

<h3c>Reset saved-configuration
<h3c>reboot

3. 项目主要实训步骤

（1）按拓扑图连接设备，并完成实训的基本配置（包括 IP 地址配置，设备命名）。

①RTA 的配置。

```
[H3C]sysname RTA
[RTA]dis ip int b
*down: administratively down
(s): spoofing
Interface              Physical Protocol IP Address    Description
Aux0                   down     down     unassigned    Aux0 Inte...
Ethernet0/0            up       down     unassigned    Ethernet0...
Ethernet0/1            down     down     unassigned    Ethernet0...
Serial1/0              down     down     unassigned    Serial1/0...
Serial2/0              down     down     unassigned    Serial2/0...
[RTA]int e0/0
[RTA-Ethernet0/0]ip address 202.1.1.2 24
[RTA-Ethernet0/0]
%Aug  8 13:15:03:510 2010 RTA IFNET/4/UPDOWN:
   Line protocol on the interface Ethernet0/0 is UP
[RTA-Ethernet0/0]
```

②RTB 的配置。

```
[H3C]sysname RTB
[RTB]dis ip int b
*down: administratively down
(s): spoofing
Interface              Physical Protocol IP Address    Description
Aux0                   down     down     unassigned    Aux0 Inte...
Ethernet0/0            up       up       unassigned    Ethernet0...
Ethernet0/1            down     down     unassigned    Ethernet0...
Serial1/0              up       up       unassigned    Serial1/0...
Serial2/0              down     down     unassigned    Serial2/0...
[RTB]int e0/0
[RTB-Ethernet0/0]ip address 202.1.1.1 24
[RTB-Ethernet0/0]
%Feb 28 13:28:06:111 2009 RTB IFNET/4/UPDOWN:
   Line protocol on the interface Ethernet0/0 is UP
[RTB-Ethernet0/0]int s1/0
[RTB-Serial1/0]ip address 192.1.0.1 24
```

③RTC 的配置。

[H3C]sysname RTC
[RTC]dis ip int b
*down: administratively down
(s): spoofing

Interface	Physical	Protocol	IP Address	Description
Aux0	down	down	unassigned	Aux0 Inte...
Ethernet0/0	up	down	unassigned	Ethernet0...
Ethernet0/1	up	down	unassigned	Ethernet0...
Serial1/0	up	up	unassigned	Serial1/0...
Serial2/0	down	down	unassigned	Serial2/0...

[RTC]int s1/0
[RTC-Serial1/0]ip address 192.1.0.2 24
[RTC-Serial1/0]
%Aug   8 13:17:41:717 2010 H3C IFNET/4/UPDOWN:
   Protocol PPP IPCP on the interface Serial1/0 is UP
[RTC-Serial1/0]int e0/0
[RTC-Ethernet0/0]ip address 192.1.1.1 24
[RTC-Ethernet0/0]
%Aug   8 13:18:24:679 2010 H3C IFNET/4/UPDOWN:
   Line protocol on the interface Ethernet0/0 is UP
[RTC-Ethernet0/0]int e0/1
[RTC-Ethernet0/1]ip address 192.1.2.1 24
[RTC-Ethernet0/1]
%Aug   8 13:19:27:904 2010 H3C IFNET/4/UPDOWN:
   Line protocol on the interface Ethernet0/1 is UP
[RTC]int loopback1
[RTC-LoopBack1]ip address 192.1.3.1 32

(2)对各路由器分别进行配置如下：

①RTA 的配置。

[RTA]rip
[RTA-rip-1]undo summary
[RTA-rip-1]network 202.1.1.0
[RTA-rip-1]quit
[RTA]int e0/0
[RTA-Ethernet0/0]rip version 2

②RTB 的配置。

[RTB]rip
[RTB-rip-1]undo summary
[RTB-rip-1]network 202.1.1.0
[RTB-rip-1]network 192.1.0.0
[RTB-rip-1]version 2
[RTB-rip-1]version 2
[RTB-rip-1]quit

[RTB]int e0/0
[RTB-Ethernet0/0]rip version 2
[RTB-Ethernet0/0]int s1/0
[RTB-Serial1/0]rip version 2
[RTB-Serial1/0]quit
[RTB]ip ip-prefix p1 permit 192.1.1.0 24
[RTB]ip ip-prefix p1 permit 192.1.2.0 24
[RTB]rip
[RTB-rip-1]filter-policy ip-prefix p1 export
[RTB-rip-1]quit
[RTB]

③RTC 的配置。

[RTC]rip
[RTC-rip-1]undo summary
[RTC-rip-1]network 192.1.0.2
[RTC-rip-1]network 192.1.1.0
[RTC-rip-1]network 192.1.2.0
[RTC-rip-1]network 192.1.3.0
[RTC-rip-1]version 2
[RTC-rip-1]quit
[RTC]int e0/1
[RTC-Ethernet0/1]rip version 2
[RTC-Ethernet0/1]int e0/0
[RTC-Ethernet0/0]rip version 2
[RTC-Ethernet0/0]int s1/0
[RTC-Serial1/0]rip version 2
[RTC-Serial1/0]int l1
[RTC-LoopBack1]rip version 2
[RTC-LoopBack1]quit
[RTC]

（3）观察现象。

在 RTB 上观察 RIP 路由信息的交互过程：

<RTB>
  *Feb 28 13:53:57:735 2009 RTB RM/6/RMDEBUG: RIP 1 : Receive response from 192.1.0.2 on Serial1/0
  *Feb 28 13:53:57:735 2009 RTB RM/6/RMDEBUG: Packet : vers 2, cmd response, length 64
  *Feb 28 13:53:57:736 2009 RTB RM/6/RMDEBUG: AFI 2, dest 192.1.1.0/255.255.255.0, nexthop 0.0.0.0, cost 1, tag 0
  *Feb 28 13:53:57:736 2009 RTB RM/6/RMDEBUG: AFI 2, dest 192.1.2.0/255.255.255.0, nexthop 0.0.0.0, cost 1, tag 0
  *Feb 28 13:53:57:736 2009 RTB RM/6/RMDEBUG: AFI 2, dest 192.1.3.1/255.255.255.255, nexthop 0.0.0.0, cost 1, tag 0
<RTB>
<RTB>
  *Feb 28 13:54:24:71 2009 RTB RM/6/RMDEBUG: RIP 1 : Sending response on interface Ethernet0/0

from 202.1.1.1 to 224.0.0.9
　　　*Feb 28 13:54:24:71 2009 RTB RM/6/RMDEBUG:　　Packet : vers 2, cmd response, length 44
　　　*Feb 28 13:54:24:71 2009 RTB RM/6/RMDEBUG:　　AFI 2, dest 192.1.1.0/255.255.255.0, nexthop 0.0.0.0, cost 2, tag 0
　　　*Feb 28 13:54:24:71 2009 RTB RM/6/RMDEBUG:　　AFI 2, dest 192.1.2.0/255.255.255.0, nexthop 0.0.0.0, cost 2, tag 0

可见，在 RIP 信息交互的过程中，发现 RTB 已经接受了去往 192.1.1.0、192.1.2.0 和 192.1.3.0 网段的路由信息，但在发送过程中仅仅发送了去往 192.1.1.1 和 192.1.2.0 网段的路由信息，说明在 RTB 上的路由策略已经生效，将去往 192.1.3.0 的路由信息过滤掉，不发送给 RTA。

25.4　项目总结与提高

写出主要项目实施规划、步骤与实训所得的主要结论。

项目二十六　VRRP 协议

26.1　项目提出

小王在某公司上班，想了解 VRRP 协议，于是他设计了如下实验。

26.2　项目分析

1．项目实训目的

掌握 VRRP 的基本命令；
理解 VRRP 的基本原理。

2．项目实现功能

通过实训，掌握 VRRP 的基本命令，理解 VRRP 的基本原理。

3．项目主要应用的技术介绍

VRRP（Virtual Router Redundancy Protocol，虚拟路由器冗余协议）将可以承担网关功能的一组路由器加入到备份组中，形成一台虚拟路由器，由 VRRP 的选举机制决定哪台路由器承担转发任务，局域网内的主机只须将虚拟路由器配置为缺省网关。

VRRP 是一种路由容错协议，也可以叫做备份路由协议。一个局域网络内的所有主机都设置缺省路由（网关），当网内的主机发出的目的地址不在本网段时，报文将通过缺省路由发往外部路由器，从而实现主机与外部网络的通信。当缺省路由器 down 掉(即端口关闭)之后，内部主机将无法与外部进行通信，如果设置了 VRRP，这时虚拟路由将启用备份路由器，从而实现可靠的通信。

26.3　项目实施

1．项目拓扑图

VRRP 协议如图 26-1 所示。

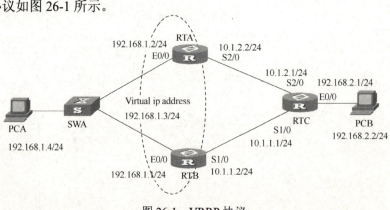

图 26-1　VRRP 协议

2. 项目实训环境准备

MSR20-40（3 台），交换机（1 台），计算机（2 台）、网线若干条。CMW 版本：5.20
为了不受原来的配置影响，在实训之前先将所有的配置数据擦除后重新启动，命令为：

    \<h3c>Reset saved-configuration
    \<h3c>reboot

3. 项目主要实训步骤

实训任务一：VRRP 单备份组配置

（1）按照实训拓扑图连接设备，并给设备命名和 IP 地址的配置。

①RTA 的配置。

```
[H3C]sysname RTA
[RTA]dis ip int b
*down: administratively down
(s): spoofing
Interface              Physical Protocol IP Address    Description
Aux0                   down     down     unassigned    Aux0 Inte...
Ethernet0/0            up       down     unassigned    Ethernet0...
Ethernet0/1            down     down     unassigned    Ethernet0...
Serial1/0              down     down     unassigned    Serial1/0...
Serial2/0              up       up       unassigned    Serial2/0...
[RTA]int e0/0
[RTA-Ethernet0/0]ip address 192.168.1.2 24
[RTA-Ethernet0/0]int s2/0
[RTA-Serial2/0]ip address 10.1.2.2 24]
```

②RTB 的配置。

```
[H3C]sysn RTB
[RTB]dis ip int b
*down: administratively down
(s): spoofing
Interface              Physical Protocol IP Address    Description
Aux0                   down     down     unassigned    Aux0 Inte...
Ethernet0/0            up       down     unassigned    Ethernet0...
Ethernet0/1            down     down     unassigned    Ethernet0...
Serial1/0              up       up       unassigned    Serial1/0...
Serial2/0              down     down     unassigned    Serial2/0...
[RTB]int e0/0
[RTB-Ethernet0/0]ip add 192.168.1.1 24
[RTB-Ethernet0/0]
%Feb 28 15:34:48:418 2009 RTB IFNET/4/UPDOWN:
   Line protocol on the interface Ethernet0/0 is UP
[RTB-Ethernet0/0]int s1/0
[RTB-Serial1/0]ip address 10.1.1.2 24
[RTB-Serial1/0]quit
```

③RTC 的配置。

```
[H3C]sysname RTC
[RTC]dis ip int b
*down: administratively down
(s): spoofing
Interface                Physical Protocol IP Address    Description
Aux0                     down     down     unassigned    Aux0 Inte...
Ethernet0/0              up       down     unassigned    Ethernet0...
Ethernet0/1              down     down     unassigned    Ethernet0...
Serial1/0                up       up       unassigned    Serial1/0...
Serial2/0                up       up       unassigned    Serial2/0...
[RTC]int s1/0
[RTC-Serial1/0]ip address 10.1.1.1 24
[RTC-Serial1/0]
%Aug   8 15:24:20:984 2010 RTC IFNET/4/UPDOWN:
   Protocol PPP IPCP on the interface Serial1/0 is UP
[RTC-Serial1/0]int s2/0
[RTC-Serial2/0]ip address 10.1.2.1 24
[RTC-Serial2/0]
%Aug   8 15:24:46:791 2010 RTC IFNET/4/UPDOWN:
   Protocol PPP IPCP on the interface Serial2/0 is UP
[RTC-Serial2/0]
[RTC-Serial2/0]int e0/0
[RTC-Ethernet0/0]ip address 192.168.2.1 24
[RTC-Ethernet0/0]quit
[RTC]
```

④依据实训组网图的标识完成 RTA、RTB、RTC、PCA、PCB 的 IP 地址配置，要实现 VRRP 备份的要求，那么 PCA 的网关地址应设置为 192.168.1.3，PCB 的网关地址应设置为 192.168.2.1。

（2）为了配置简单，在 RTA、RTB、RTC 上运行 OSPF，所有接口网段都在 OSPF area 0 中发布。

①RTA 的配置。

```
[RTA]ospf
[RTA-ospf-1]area 0
[RTA-ospf-1-area-0.0.0.0]network 192.168.1.0 0.0.0.255
[RTA-ospf-1-area-0.0.0.0]network 10.1.2.0 0.0.0.255
[RTA-ospf-1-area-0.0.0.0]quit
[RTA-ospf-1]quit
[RTA]
```

②RTB 的配置。

```
[RTB]ospf 1
[RTB-ospf-1]area 0
[RTB-ospf-1-area-0.0.0.0]network 192.168.1.0 0.0.0.255
```

```
[RTB-ospf-1-area-0.0.0.0]network 10.1.1.0 0.0.0.255
[RTB-ospf-1-area-0.0.0.0]quit
[RTB-ospf-1]quit
[RTB]
```

③RTC 的配置。

```
[RTC]ospf 1
[RTC-ospf-1]area 0
[RTC-ospf-1-area-0.0.0.0]network 10.1.2.0 0.0.0.255
[RTC-ospf-1-area-0.0.0.0]network 10.1.1.0 0.0.0.255
[RTC-ospf-1-area-0.0.0.0]network 192.168.2.0 0.0.0.255
[RTC-ospf-1-area-0.0.0.0]quit
[RTC-ospf-1]quit
[RTC]
[RTC]dis ip rou
Routing Tables: Public
        Destinations : 11       Routes : 12
Destination/Mask    Proto   Pre   Cost      NextHop         Interface
10.1.1.0/24         Direct  0     0         10.1.1.1        S1/0
10.1.1.1/32         Direct  0     0         127.0.0.1       InLoop0
10.1.1.2/32         Direct  0     0         10.1.1.2        S1/0
10.1.2.0/24         Direct  0     0         10.1.2.1        S2/0
10.1.2.1/32         Direct  0     0         127.0.0.1       InLoop0
10.1.2.2/32         Direct  0     0         10.1.2.2        S2/0
127.0.0.0/8         Direct  0     0         127.0.0.1       InLoop0
127.0.0.1/32        Direct  0     0         127.0.0.1       InLoop0
192.168.1.0/24      OSPF    10    1563      10.1.2.2        S2/0
                    OSPF    10    1563      10.1.1.2        S1/0
192.168.2.0/24      Direct  0     0         192.168.2.1     Eth0/0
192.168.2.1/32      Direct  0     0         127.0.0.1       InLoop0
```

（3）配置 VRRP。

创建 VRRP 备份组的同时，需要在接口视图下配置备份组的虚拟 IP 地址，并且保证配置的虚拟 IP 地址与 RTA 和 RTB E0/0 接口的 IP 地址在同一网段。

①配置 RTA 虚拟 IP 地址：

```
[RTA-Ethernet0/0]vrrp vrid 1 virtual-ip 192.168.1.3
[RTB-Ethernet0/0]vrrp vrid 1 virtual-ip 192.168.1.3
```

从 PCA ping 自己的虚拟网关，显示能够 ping 通。

②接下来配置备份组优先级以确保在初始情况下，RTA 为 Master 路由器承担业务转发，在 MSR 路由器上 VRRP 备份组的缺省优先级是 100，要确保 RTA 为 VRRP 备份组 Master 路由器，那么 RTA 在该备份组中的优先级应该大于（大于/小于）RTB 在该备份组中的优先级。

配置 RTA 备份组优先级为 120：

```
[RTA- GigabitEthernet0/0] vrrp vrid 1 priority 120
```

配置 RTB 备份组优先级为 100：

[RTB- GigabitEthernet0/0] vrrp vrid 1 priority 100

RTB 不需此配置也可以，缺省的备份组优先级就是 100。

③验证 VRRP：

在 PCA 上用 ping 检测到 PCB 的可达性，其结果是可以 ping 通。

在 RTA 上通过命令 display vrrp 查看 VRRP 备份组状态的摘要信息，通过命令 display vrrp verbose 可以查看 VRRP 备份组状态的详细信息。

```
[RTA]display vrrp verbose
 IPv4 Standby Information:
 Run Method         : VIRTUAL-MAC
 Total number of virtual routers: 1
 Interface          : Ethernet0/0
   VRID             : 1              Adver. Timer   : 1
   Admin Status     : UP             State          : Master
   Config Pri       : 120            Run Pri        : 120
   Preempt Mode     : YES            Delay Time     : 0
   Auth Type        : NONE
   Virtual IP       : 192.168.1.3
   Virtual MAC      : 0000-5e00-0101
   Master IP        : 192.168.1.2
```

根据该命令输出，可以看出 RTA 的 VRRP 状态是 Master。在 RTB 上执行同样的命令。

```
[RTB]display vrrp verbose
 IPv4 Standby Information:
 Run Method         : VIRTUAL-MAC
 Total number of virtual routers: 1
 Interface          : Ethernet0/0
   VRID             : 1              Adver. Timer   : 1
   Admin Status     : UP             State          : Backup
   Config Pri       : 100            Run Pri        : 100
   Preempt Mode     : YES            Delay Time     : 0
   Auth Type        : NONE
   Virtual IP       : 192.168.1.3
   Master IP        : 192.168.1.2
```

可以看到 RTB 的 VRRP 状态是 Backup。

此时将 RTA 关机，再次在 PCA 上用 ping 检测到 PCB 的可达性，其结果是依然可以 ping 通，此时在 RTB 上查看 VRRP 状态。

```
[RTB]display vrrp verbose
 IPv4 Standby Information:
 Run Method         : VIRTUAL-MAC
 Total number of virtual routers: 1
 Interface          : Ethernet0/0
   VRID             : 1              Adver. Timer   : 1
   Admin Status     : UP             State          : Master
   Config Pri       : 100            Run Pri        : 100
```

Preempt Mode	: YES	Delay Time	: 0
Auth Type	: NONE		
Virtual IP	: 192.168.1.3		
Virtual MAC	: 0000-5e00-0101		
Master IP	: 192.168.1.1		

可以看到 RTB 的 VRRP 状态是 Master。

实训任务二：VRRP 监视接口配置

打开 RTA 电源。

（1）配置 VRRP 指定被监视的接口。

①在 RTA 和 RTB 上的 Ethernet0/0 接口下配置 VRRP 监视上行出口 Serial1/0，当上行出口 Serail 1/0 出现故障时，路由器的优先级自动降低 30，以低于处于备份组的路由器优先级，从而实现主备倒换。请在下面的空格中补充完整的命令。

配置 RTA：

　　[RTA-Ethernet0/0] vrrp vrid 1 track interface Serial1/0 reduced 30

配置 RTB：

　　[RTB-Ethernet0/0] vrrp vrid 1 track interface Serial2/0 reduced 30

②在 RTA、RTB 上都做了如下的配置。

　　[RTA-Ethernet0/0] vrrp vrid 1 timer advertise 5
　　[RTB-Ethernet0/0] vrrp vrid 1 timer advertise 5

该配置命令的含义是设置备份组中的 Master 路由器发送 VRRP 通告报文的时间间隔为 5 秒。

③配置备份组中的路由器工作在抢占方式，并配置抢占延迟时间为 5 秒。

　　[RTA-Ethernet0/0] vrrp vrid 1 preempt-mode timer delay 5
　　[RTB-Ethernet0/0] vrrp vrid 1 preempt-mode timer delay 5

（2）验证 VRRP。

①在 PCA 上用 ping 检测到 PCB 的可达性，其结果是可以 ping 通。（略）

②在 RTA 上通过命令 display vrrp verbose 查看 VRRP 备份组状态的摘要信息。

[RTA]dis vrrp verbose
IPv4 Standby Information:
Run Method　　　　: VIRTUAL-MAC
Total number of virtual routers: 1

Interface	: Ethernet0/0		
VRID	: 1	Adver. Timer	: 5
Admin Status	: UP	State	: Master
Config Pri	: 120	Run Pri	: 120
Preempt Mode	: YES	Delay Time	: 5
Auth Type	: NONE		
Track IF	: S2/0	Pri Reduced	: 30
Virtual IP	: 192.168.1.3		
Virtual MAC	: 0000-5e00-0101		

```
    Master IP              : 192.168.1.2
```

通过命令 display vrrp verbose 可以查看 VRRP 备份组状态的详细信息，根据该命令输出，可以看出 RTA 的 VRRP 状态是 Maste，路由器优先级是 120。

③在 RTB 上执行同样的命令。

```
[RTB]display vrrp verbose
 IPv4 Standby Information:
 Run Method            : VIRTUAL-MAC
 Total number of virtual routers: 1
 Interface             : Ethernet0/0
 VRID                  : 1              Adver. Timer    : 5
 Admin Status          : UP             State           : Backup
 Config Pri            : 100            Run Pri         : 100
 Preempt Mode          : YES            Delay Time      : 5
 Auth Type             : NONE
 Track IF              : S1/0           Pri Reduced     : 30
 Virtual IP            : 192.168.1.3
 Master IP             : 192.168.1.2
```

可以看到 RTB 的 VRRP 状态是 BACKUP，路由器优先级是 100。

④此时将 RTA 连接 RTC 的接口 Serail 2/0 shutdown。

```
[RTA-Serial2/0]shutdown
```

再次在 PCA 上用 ping 检测到 PCB 的可达性，其结果是依然可以 ping 通，此时在 RTA 上查看 VRRP 状态。

```
[RTA-Serial2/0]quit
[RTA]display vrrp verbose
 IPv4 Standby Information:
 Run Method            : VIRTUAL-MAC
 Total number of virtual routers: 1
 Interface             : Ethernet0/0
 VRID                  : 1              Adver. Timer    : 5
 Admin Status          : UP             State           : Backup
 Config Pri            : 120            Run Pri         : 90
 Preempt Mode          : YES            Delay Time      : 5
 Auth Type             : NONE
 Track IF              : S2/0           Pri Reduced     : 30
 Virtual IP            : 192.168.1.3
 Master IP             : 192.168.1.1
```

可以看到 RTA 的 VRRP 状态是 Backup，路由器优先级是 90。

⑤在 RTB 上查看 VRRP 状态。

```
[RTB]dis vrrp verbose
 IPv4 Standby Information:
 Run Method            : VIRTUAL-MAC
 Total number of virtual routers: 1
```

```
Interface         : Ethernet0/0
VRID              : 1                    Adver. Timer   : 5
Admin Status      : UP                   State          : Master
Config Pri        : 100                  Run Pri        : 100
Preempt Mode      : YES                  Delay Time     : 5
Auth Type         : NONE
Track IF          : S1/0                 Pri Reduced    : 30
Virtual IP        : 192.168.1.3
Virtual MAC       : 0000-5e00-0101
Master IP         : 192.168.1.1
```

可以看到 RTB 的 VRRP 状态是 Master，路由器优先级是 100。从如上显示信息可以看出，由于上行接口 Serail 1/0 出现故障，VRRP 备份组进行了主备倒换。

26.4　项目总结与提高

写出主要项目实施规划、步骤与实训所得的主要结论。

实训报告的基本内容及要求

每门课程的所有实训项目的报告必须以课程为单位装订成册，格式参见附件1。

实训报告应体现预习、实训记录和实训报告，要求这三个过程在一个实训报告中完成。

1. 实训预习

在实训前每位同学都需要对本次实训进行认真的预习，并写好预习报告，在预习报告中要写出实训目的、要求，需要用到的仪器设备、物品资料以及简要的实训步骤，形成一个操作提纲。对实训中的安全注意事项及可能出现的现象等做到心中有数，但这些不要求写在预习报告中。

设计性实训要求进入实训室前写出实训方案。

2. 实训记录

学生开始实训时，应该将记录本放在旁边，将实训中所做的每一步操作、观察到的现象和所测得的数据及相关条件如实地记录下来。

实训记录中应有指导教师的签名。

3. 实训总结

主要内容包括对实训数据、实训中的特殊现象、实训操作的成败、实训的关键点等内容进行整理、解释、分析总结，回答思考题，提出实训结论或提出自己的看法等。